Improving Energy Demand Analysis

Paul C. Stern, Editor

Panel on Energy Demand Analysis

Committee on Behavioral and Social
Aspects of Energy Consumption
and Production

Commission on Behavioral and Social
Sciences and Education

National Research Council

NATIONAL ACADEMY PRESS
Washington, D.C.   1984

NATIONAL ACADEMY PRESS ● 2101 Constitution Avenue, N.W. ● Washington, D.C. 20418

Library of Congress Catalog Card Number 84-60902
International Standard Book Number 0-309-03477-9

Printed in the United States of America

Panel on Energy Demand Analysis

JOHN M. DARLEY (Chair), Department of Psychology,
    Princeton University
DAVID A. FREEDMAN, Department of Statistics, University
    of California, Berkeley
DANIEL H. HILL, Institute for Social Research, University
    of Michigan Ann Arbor
ERIC HIRST, Energy Division, Oak Ridge National
    Laboratory, Oak Ridge, Tennessee
DANIEL L. MCFADDEN, Department of Economics,
    Massachusetts Institute of Technology
LINCOLN MOSES, Department of Statistics, Stanford
    University

PAUL C. STERN, Study Director

Committee on Behavioral and Social Aspects of Energy Consumption and Production

ELLIOT ARONSON (Chair), Stevenson College, University of
   California, Santa Cruz
ROBERT AXELROD, Institute of Public Policy Studies,
   University of Michigan
JOHN M. DARLEY, Department of Psychology, Princeton
   University
SARA B. KIESLER, Department of Social Science,
   Carnegie-Mellon University
DOROTHY LEONARD-BARTON, School of Business, Harvard
   University
JAMES G. MARCH, Graduate School of Business, Stanford
   University
JAMES N. MORGAN, Survey Research Center, University of
   Michigan
PETER MORRISON, The Rand Corporation, Santa Monica,
   California
LINCOLN MOSES, Department of Statistics, Stanford
   University
LAURA NADER, Department of Anthropology, University of
   California, Berkeley
STEVEN E. PERMUT, School of Organization and Management,
   Yale University
ALLAN SCHNAIBERG, Department of Sociology, Northwestern
   University
ROBERT H. SOCOLOW, Center for Energy and Environmental
   Studies, Princeton University
THOMAS J. WILBANKS, Energy Division, Oak Ridge National
   Laboratory, Oak Ridge, Tennessee
SIDNEY WINTER, School of Organization and Management,
   Yale University

Preface

Since 1980 a Committee on Behavioral and Social Aspects of Energy Consumption and Production has been functioning within the National Research Council to draw on knowledge from the social and behavioral sciences to improve understanding of energy consumption and production in the United States. That committee's report (Stern and Aronson, 1984) developed a new perspective on energy issues and applied it to three areas of energy policy: energy use and conservation, energy emergencies, and energy activity at the local level.

The committee's behavioral perspective on energy use was very different from that usually used in energy demand models; in fact, the committee's report directly questioned the typical assumption that energy users can be adequately characterized as rational economic actors making choices in a market. In light of this report, as well as some evidence that demand models had not been functioning adequately to forecast changes in energy markets, the Office of Policy, Planning, and Analysis of the U. S. Department of Energy called upon the committee in 1983 to further examine the behavioral factors identified in its earlier work "so as to inform policy analysis and enhance current analytical techniques, especially energy policy models." The charge was to explore questions such as the following:

1. Do consumers respond to levels of energy prices, changes in price, or the rate of change in price? In other words, are responses to "price shocks" different from ordinary price responses?
2. How is the price elasticity of demand affected by other dynamically changing forces, such as the quality of energy information available to producers and consumers?

3. Is a secular change beginning in consumer preferences that will affect energy demand over the long term independently of market events?
4. What might be useful early indicators of secular trends in consumer preference, interactions between energy prices and other variables, or other postulated influences on energy demand?
5. How could future projections of energy demand take issues such as the above into account?

The Department of Energy recognized that it would be impossible to answer all these questions on the basis of existing knowledge; the hope was rather to build a bridge between the analyses of social and behavioral scientists on one side and energy demand modelers on the other, and to thus improve communication between the groups in order to develop better energy analyses.

Examining the fields of expertise needed for the task led the committee to the decision that it would be most appropriate to create a new group, the Panel on Energy Demand Analysis. The panel's members vary widely in perspective and training. Some have studied energy demand from the perspective of economics and others from the viewpoints of other behavioral sciences. Some have been generally sympathetic to the modeling enterprises and others have been critical. In spite of—or perhaps because of—such differences, the panel rapidly converged on a shared definition of its work.

Since the time frame for the report was quite constrained and the resources available were quite limited (insufficient, for example, to commission comprehensive literature reviews, large-scale data collection projects, or complete analyses of demand models in use for energy demand predictions), the essential task of the panel had to be a more general and abstract analysis of various analytic strategies for predicting energy demand and evaluating policy options. As would be expected, given the constraints, the panel often raised questions without attempting to provide the answers. One of the major conclusions of our report is that extensive data collection efforts and theoretical development are necessary before it will be possible to make adequate demand forecasts or thorough and reliable policy analyses.

Although this report is a document of the whole panel, the chapters reflect a division of labor in writing. Chapter 1 was drafted by Paul Stern; Chapter 2 was drafted by Daniel Hill; Chapter 3 was a collaboration of Paul

Stern, Linda Berry of Oak Ridge National Laboratory, and Eric Hirst; Chapter 4 was drafted by John Darley and Paul Stern; Chapter 5 was drafted by Daniel McFadden; and Chapter 6 was a group product that Paul Stern put into its final form. As has already been mentioned, the report was produced in what, at least to an academic, seems to be a remarkably short period of time and while the panel members were heavily engaged in their regular patterns of activities. For these reasons, as well as his own skills and knowledge in the area, a considerable share of the final writing and editing of the report was done by Paul Stern, the panel's study director.

Helpful comments were received from James Morgan and Thomas Wilbanks of the parent committee and from reviewers for the Commission on Behavioral and Social Sciences and Education. In addition, it is a pleasure to express appreciation to several other people who made important contributions to the panel's work. These include John Corliss, John Stanley-Miller, and Barry McNutt of the Department of Energy; David Goslin and Heidi Hartmann, executive director and associate executive director of the Commission on Behavioral and Social Sciences and Education; and Eugenia Grohman, the Commission's associate director for reports.

JOHN M. DARLEY, Chair
Panel on Energy Demand Analysis

Contents

Improving Energy Demand Analysis

1

Formal Modeling and Problem-Oriented Research

UNDERSTANDING THE DEMAND FOR ENERGY

The past decade has demonstrated the need to improve
understanding of U.S. energy demand. Demand projections
have repeatedly proved inaccurate--usually on the high
side--and there is little reason for confidence that
today's projections will be any more accurate than those
of the past. Yet accurate analysis of energy demand
remains an important national priority. It is needed if
policy makers are to find effective ways to avoid national
crises when there are shocks in world oil markets. It is
needed to set appropriate levels of investment in electric
power generation so that the country can avoid the economic drain of unnecessary investment and the threat of
widespread brownouts and blackouts. And it is needed to
aid in decisions about how serious a public effort is
needed in the areas of energy production and conservation.
But in spite of this clear need, it has so far proved
impossible to differentiate accurate from inaccurate projections on the basis of their internal characteristics.
For example, projections from detailed formal models of
the energy system have often been as far wrong as the
intuitive judgments of experts or forecasts from simpler
models (e.g., Ascher, 1978; McNees, 1979; Zarnowitz,
1979).

There are at least two inherent problems in trying to
analyze energy demand for policy purposes. One problem
is that major changes in the energy environment seem
likely in the near future, changes similar in magnitude
to those produced by the 1973 oil embargo and the 1979
revolution in Iran. It is unreasonable to expect energy
analyses to predict such events because they stem from
political and social forces outside the energy system:

1

they are, in a fundamental sense, not energy events.
Because these events impinge on energy systems from outside, even models based on accurate theory about the
energy system will be wrong much of the time.

The second inherent problem is that good energy models
are few and far between. The dynamics of energy demand
are far from well understood, even barring political cata-
clysms. Economic theory describes some general depen-
dencies of energy demand on income, output, and the costs
of factors of production, but it leaves great latitude for
the behavior of energy users, both individually and in the
aggregate. When energy prices rise, for example, other
factors of production (labor or capital, for example) tend
to be substituted, but the degree of substitution and its
path over time can only be determined empirically. This
kind of empirical work is hard to do without a solid
foundation of theory, method, and data. Direct observation is difficult and yields only imprecise estimates
because the behavior occurs in a complex and rapidly
changing environment.

The difficulties of modeling demand are general, with
energy a representative and important case. Simple economic models, even those that track past experience well
in the aggregate, do not completely describe the bases of
consumers' responses. Such models treat consumers in
general--and energy users in particular--as rational economic actors operating in an environment roughly describ-
able as a market. But people often fail to take actions
that would offer them clear economic benefits (e.g.,
Office of Technology Assessment, 1982; Solar Energy
Research Institute, 1981; Stobaugh and Yergin, 1979).
Furthermore, their investments in energy efficiency can
be affected by a range of variables that are rarely estimated in energy demand models. Several of these variables
are outlined below; they are discussed in more detail in
the report of the Committee on Behavioral and Social
Aspects of Energy Consumption and Production (Stern and
Aronson, 1984).

Preferences Consumer preferences cannot always be reduced
to a common metric. People sometimes invest in energy
efficiency to save money, sometimes to gain such intangi-
ble consumer benefits as freedom from drafts in houses.
Ways of incorporating such preferences in broad policy
analyses have not yet been developed. Furthermore, claims
of a rise of values of voluntary simplicity (Elgin, 1981)
or spirituality (Yankelovich, 1982) suggest that past

relationships between increased income or gross national product (GNP) and energy demand may soon become weaker. Not enough is known to evaluate such hypotheses.

Institutional Limits on Choice Energy demand is affected by such institutional factors as regulatory constraints on energy prices; the split incentives that operate when the owner and the user of energy-using equipment and buildings are not the same (e.g., rental housing); and the power of manufacturers, standard-setting groups, lenders, and professional organizations to limit the range of investment alternatives available to energy users. Information is scanty, however, on how much influence each of these factors has and on the way each might respond to changes in policy or in the energy environment. As a result, policy analyses are often weak when they address institutional factors.

Attitudes, Values, and Beliefs A number of survey and experimental studies show that various values, attitudes, and beliefs mediate the effects of structural, demo-graphic, and economic variables on energy-using behavior. Sometimes they appear to affect energy use more strongly than price does (e.g., Heberlein and Warriner, 1982). Stern and Oskamp (1984) have summarized this literature and offer a theoretical account that describes the joint effect on demand of economic, structural, demographic, and psychological variables. This framework may prove ana-lytically useful; it certainly underlines the extent of missing knowledge about energy demand. As yet, little is known about the conditions under which personal commit-ment, normative beliefs, or other attitudes or values make an important difference in energy demand. And even the rather detailed model of Stern and Oskamp does not address such potentially important variables as expectations of future prices and availability of fuels, beliefs about uncertainty of the same, willingness to seek and process information, and the existence of government or utility conservation programs.

Communication There is considerable evidence that consumers' energy choices can be influenced in important ways by communication through informal social networks and by levels of trust in available information (for a review, see Stern and Aronson, 1984:Chapter 4). These social influences have many policy implications, but little thought has been given to incorporating them in general analyses of energy demand.

Convenience Economic theory recognizes the importance of the cost of gathering information and of uncertainty, and there is evidence that efforts to reduce such cost and uncertainty affect the choices of energy users. For example, some residential energy conservation programs seem to be successful partly because they simplify decisions and protect people from incompetent or unscrupulous contractors (Stern, Black, and Elworth, 1981). But the convenience factor has not been incorporated in energy models.

Symbolic Meanings Energy users probably respond to government programs for conservation as a function of what conservation has come to mean to them in a public policy context (Stern and Aronson, 1984). Thus, President Carter's famous energy speech in his sweater symbolized an identity between conservation and being uncomfortably cold and reinforced this connection in people's minds. The effects of such symbolism, however real, are rarely considered explicitly in energy analyses.

The Psychology of Shortage Some commentators say that public awareness about energy was changed in 1979 in a way that did not happen in 1973, even though in both periods there were rapid price increases for oil and the threat of serious shortage. Some hypothesize that it took a second "shock" to convince people that the energy environment had changed. There is as yet no accepted way of evaluating the existence or quantifying the importance of this psychology of shortage.

To the extent that factors such as those noted above determine behavior, the principles that govern the behavior of energy users do not lead to economically rational action as this concept is realized in existing models of energy demand.[1] Thus, the most readily

---

[1]Whether all the relevant nonmonetary variables are measurable in principle and whether the behavior that flows from them can all be construed as rational under some noncircular definition of utility are two issues we do not address. It is clear, however, that some nonmonetary variables not now addressed in economic models

available analytical framework is almost certaintly incomplete and may even be inaccurate or misleading. Formal models based on the assumption of rational choice may not capture the appropriate variables for analyzing energy users' responses to changing conditions and may, as a result, be quite wrong about the level of short-run response to changes in the energy environment and about the ultimate penetration of energy-efficient technology. Furthermore, an exclusive focus on the variables included in energy models implies neglect of other factors, some of which may be controllable by policy.

A more complete framework for analyzing energy demand, suitable for general application and specified in quantitative form, is not now available, but the concepts and findings of behavioral research show promise of improving this situation. Behavioral research on energy users and their environment, including small-scale controlled experiments, can probably improve understanding of energy use in ways that will be useful for policy-related demand analyses, including formal energy modeling, and that may have policy applications. For example, laboratory experiments with different energy-efficiency labels for appliances can help determine the features of energy information that make the most difference in appliance purchase decisions. Such knowledge would be useful both for modeling those decisions and for designing effective appliance labels.

ABOUT THIS BOOK

This book examines the state of energy demand analysis and develops recommendations for improving it. A recurring theme is the limitations of formal modeling--the dominant form of demand analysis. Despite the value of modeling, serious gaps in theory and empirical knowledge have led existing models to systematically overlook important factors affecting energy demand, such as those noted above. We conclude that both energy demand analysis and the policy choices it supports can benefit from improved data

---

do affect energy demand. It is also clear that the number of nonmonetary variables plausibly affecting demand readily outstrips the ability to measure them. As a result, an analytical framework that can shorten the list of plausible influences has obvious value.

collection and from a more balanced reliance on a variety of analytic methods. These efforts will sometimes produce results that are easily incorporated into demand models by changing the parameters assigned to the variables in models, altering the functional forms of equations, or adding variables. Sometimes, however, new findings will not be so easily modeled, and more thorough changes in analytic tools will prove justified.

This book distinguishes the formal modeling approach to energy demand analysis from what we call a problem-oriented approach. While the former employs mathematical techniques to build comprehensive descriptions of a complex energy system, the latter uses a wider range of research methods--including surveys, controlled experiments, exploratory analysis of existing data, and evaluation research--to examine narrower analytic and policy questions. The rest of this chapter explores the strengths and limitations of each approach.

Chapters 2 through 4 consider three issues the panel finds particularly important or interesting: the effects of prices on energy demand; the effects of financial incentives on investments in residential energy efficiency; and the effects of information on energy demand. In these chapters, we pay particular attention to qualitative factors that have not been given careful consideration in formal models but that behavioral research has shown may have major effects on demand. For example, we explore the possibility that consumers respond to changes in price and not only to price levels; we consider the impact of qualitative differences among types of incentives, such as tax credits, rebates, and loan subsidies; and we examine the role of word-of-mouth communication in the adoption of energy-efficient technologies. We present the available evidence on these and other topics, evaluate the likely importance of some factors, and discuss methods that can be used to assess the roles such factors can play and to address their roles more fully in demand analyses.

Chapter 5 is a case study of a particular type of behavior that affects energy demand and is in turn affected by prices, incentives, and information, as discussed in Chapters 2 through 4. Taking appliance choice as an instance of such behavior, Chapter 5 identifies gaps in knowledge that impede understanding of such choice and suggests data collection efforts that could help close these gaps.

Chapter 6 presents our conclusions and recommendations about the appropriate roles of formal models, problem-oriented research, and data collection in energy demand analysis. It concludes by discussing ways different methods of analysis can complement each other.

FORMAL MODELS OF ENERGY DEMAND

Formal policy models are analytic tools that are built to provide quantitative projections of energy demand and to provide answers to questions about how demand might respond to alternative political and economic events or to policy choices. Formal models are often used to answer such questions as: "What will be the price of natural gas after full decontrol in the United States?" "What will be the effect on the world price of oil of 4 percent annual GNP growth in the United States over the next two years?"

Much of energy policy analysis is now conducted via formal models. Two advantages of formal models over informal judgments for answering such questions are that they can ensure that all parts of the energy economy are included in the analysis and that their construction requires that the assumptions on which they are based are in quantitative form. Disagreements with a thorough and explicit analysis can be directed to particular elements of it rather than to some global expert judgment. Although models vary greatly in their levels of detail and in the amount of information they require to function, most energy models are computerized so that considerable detail can be included while it is still possible to respond quickly to forecasting or policy questions.

Formal models are the favored analytical method among many policy makers and policy analysts for reasons only loosely related to their adequacy as tools for policy analysis or forecasting.[2] Models are valued because

----

[2] It is possible to distinguish the use of formal models for hypothesis-testing through analysis of relationships in existing data from their use for forecasting and policy analysis. Generally, forecasting and policy models tend to strive for completeness in the variables covered and to make projections into the future, while hypothesis-testing models tend to be less inclusive, more tightly derived from theory, and less concerned about projections.

they can respond to a range of policy questions and
because their responses are quantitative. Furthermore,
once created, models offer quick and relatively low-cost
responses. These attributes are important to policy
makers who must make many complex choices under severe
time constraints. Models offer an added attraction for
some policy makers: the mystique of computer analysis
and economic theory gives an authoritative sense to the
output from models that is quite unrelated to the adequacy
of the models being used. For all these reasons, formal
models are likely to remain the dominant tools for policy
analysis in the area of energy demand.

Types of Models

Three frequently used approaches to building formal
energy demand models are econometrics, system dynamics,
and engineering process modeling.

<u>Econometric Models</u> Econometric demand forecasting models
typically include regression equations that estimate
demand variables from other factors such as price, income,
output, or noneconomic factors.[3] When presented in
"reduced form," each equation expresses a demand variable
as a function of other variables. In principle, the
family of regression equations is fitted to a data set
based on observation of demand at various values of other
variables included in the equations. Thus, the regression
equations take into account the structure of correlations
among the variables measured. In fact, however, judgment
can be as important as measurement: the forms of the
mathematical functions used to describe relationships
among variables are generally selected by judgment or
convention; the parameters in the equations are often
adjusted judgmentally when the initial model results seem
unreasonable; and even the data used to estimate regression equations are often imputed, presumed, or guessed
rather than measured directly. For example, the rate of
capital investment in energy efficiency over time is often

The discussion of models in this section primarily concerns models as used for projection and policy analysis.
[3] The discussion here is focused primarily on models
that are fitted to aggregate data.

estimated a priori, using an assumption that it responds
to price changes with a time lag described by a particular
mathematical function. For another example, data at the
state level are sometimes constructed by allocating values
measured at the national level according to some plausible
formula. A detailed discussion of the variety of econometric energy demand models and the theoretical and empirical issues attending them can be found in Bohi (1981).

Econometric models are limited in that they can only
include variables for which data exist or for which data
can reasonably be estimated; these may not be the most
critical variables. The limitation is also an advantage,
however, in that it imposes the discipline of tying models
to data.

Projections based on econometric models are problematic
because they assume that the pattern of correlations in
available data reflects structural relationships that will
extend reliably into the future. There is little empirical or theoretical justification for such an assumption
in most cases, and the weaker the data base for the model,
the weaker the assumption. Thus, projections from econometric models tend to enshrine as economic structure any
systematic covariation, even if it was introduced by
biased measurement or estimation; the greater the number
of values and relationships that are estimated, presumed,
or imputed, the more likely it is that the projections
will contain systematic bias. This is a particularly
important problem in energy forecasting because extrapo-
lations ignore the possibility of major shocks to the
system, such as those experienced in 1973 and 1979.

System Dynamics Models System dynamics models consist of
sets of simultaneous partial differential equations each
of which, in effect, states an assumption about the
response of one variable to the values of other variables
in the immediately previous time period. The collection
of equations defines relationships of energy demand to a
set of determining factors and, with the addition of an
initial set of values, can be used to generate values
indefinitely into the future. In most cases, in which
the causal relationships are not well understood, the
modelers' judgments can determine the models' results.
Compared with econometric models, in which causal relationships can also be postulated by the modeler, system
dynamics modelers are less likely to check their postulated functional forms and parameters against data. The
reliance of econometricians on time-series data acts as a

rein on their judgment; this sort of safeguard is not an intrinsic feature of system dynamics models.

Both econometric models and system dynamics models are typically checked by beginning with knowledge about some point in the past and demonstrating that by selecting the proper variables and coefficients, the models can generate values of energy demand close to those that have been observed from that point to the present. Once the variables are chosen, coefficients for them are chosen to make the model fit the data. In selecting the variables, however, econometric models are constrained to choose among those variables for which data exist; systems dynamics models can postulate new variables. Thus, system dynamics allows modelers great freedom to construct theory in the process of building the model. A system dynamics modeler can describe a variety of possible causal patterns and can generate richly textured accounts of the future. But because relationships can be postulated without observation, system dynamics models can easily build more and more detailed pictures from less and less well-documented knowledge.[4]

What seems crucial for validating system dynamics models is empirical support for the causal relationships postulated in the modeling process. When this cannot be provided by analysis of existing data sets, it could be provided by experimental research to test the propositions embodied in the models' equations. This sort of approach, however, has not been a part of the research program of system dynamics modeling.

Engineering Process Models Engineering process models rely on detailed data on present energy demand and focus on the determination of demand by technological choice. They account for present energy use by building up from energy-using technologies, aggregating according to cate-

---

[4]A shortage of empirical support has been a problem with the system dynamics models presently available for use in the Department of Energy's Office of Policy, Planning, and Analysis (PPA). These include FOSSIL2 (Energy and Environmental Analysis, 1980), DEMAND'81 (Backus, 1981), OECD1 (Salama, Greene, and Krantzman, 1980), and HID (Marshall, 1983). The PPA recognizes the inadequacy of these models as a basis for demand analysis and has begun an effort--including the present work--to improve the situation.

gories such as appliance type, building type, end use
(e.g., space heating, materials processing), and fuel
type. To make projections, engineering models rely on
assumptions from economic and demographic projections
about such factors as household formation, building stock
replacement, and the rate of improvement in the energy
efficiency of available technologies.

The chief strength of engineering process models in
comparison with other types of models is that they can
describe the technological trade-offs in more detail. In
practice, however, there are difficulties with such
models. For example, models built from laboratory tests
of the performance of technology are often in error as
predictors of actual performance; predictions about the
effects of building insulation may be the most extreme
example. This problem can be addressed by conducting
field studies of the technologies and validating the
models on the resulting data. Another problem is that
engineering calculations of the cost of more energy-
efficient technologies may bear little resemblance to the
price of those technologies in the market. Good data on
manufacturers' decisions about pricing simply do not
exist.

Engineering models, like other types of models, also
rely greatly on judgment. For example, they generally
lack a theoretical framework for modeling choices among
technologies. As a result, their estimates of the rate
of penetration of new technology are ultimately judgmental. The obvious solution to this problem involves conducting separate studies of purchases to provide estimates
of penetration.

The above descriptions of three model types are idealized and incomplete. Other classes of models exist (see,
e.g., Greenberger, Crenson, and Crissey, 1976, for a discussion), and hybrid models have recently been developed
that combine the features of different types of models in
an attempt to gain some of the advantages of each. For
example, econometric models of appliance choice are now
incorporating engineering data (the residential end-use
energy planning system model, Goett and McFadden, 1982).
The following comments generally apply, however, to formal models as an analytical approach.

Problems with Formal Demand Models

Serious criticisms have been increasingly raised, both by
modelers and others, about formal demand models and their

use in policy analysis. We can only give a brief account
of the criticisms here; for more detailed analyses, see,
for example, Ascher (1978); Brewer (1983); Freedman
(1981); Freedman, Rothenberg, and Sutch (1983a);
Greenberger, Crenson, and Crissey (1976).

Some technical criticisms are serious: models have
been found to be poorly documented; "overfitted," meaning
that they achieve close correspondence to past experience
by making large numbers of unsupported and sometimes
unreasonable assumptions; insufficiently explicit in their
treatment of uncertainty; and inadequately tested and
reviewed (see, e.g., Brewer, 1983; Freedman, Rothenberg,
and Sutch, 1983a).

Some of the criticisms point to the lack of empirical
basis for models' estimates of the ultimate magnitude and
the rate of change of energy use in response to significant stimuli. Key parameter values are often postulated,
and relationships are assumed to follow particular mathematical forms with little or no empirical support for the
choices. Such procedures inspire little confidence.
Unfortunately, little knowledge exists for supplying the
coefficients needed to estimate the ultimate magnitude of
change (steady-state response) or the rate of change
(dynamic response) in the terms of existing models.

Models have also been criticized for a poor track
record as predictors. This criticism has been made of
large-scale economic models generally (e.g., Ascher, 1978;
McNees, 1979; Zarnowitz, 1979) and of energy models as
special cases. The recent evidence suggests the predictive success varies widely from one energy model to
another. Some of this evidence suggests that some models
have failed in predicting energy prices and not in describing the relationship of price to demand: if actual
(realized) energy prices are substituted for the prices
assumed in those models, they more correcty forecast
energy demand up to the present (McFadden, 1983).

Some of the criticisms point to the political implications of large-scale energy modeling. Brewer (1983)
argues that models tend to have a conservative bias
because they extrapolate from past experience. He also
argues that as policy analysis comes to rely on more and
more technically complex models, the public tends to be
closed out of policy debates.

Still other criticisms emphasize the consequences of
taking energy models seriously despite their flaws.
Because of insufficient knowledge, virtually every model
contains some judgmental elements that were added to make

the model conform better to the prevailing qualitative understanding of most energy modelers. These judgments are buried in models that appear superficially to be well grounded in science, and they interact with other elements of the models in unknown ways. When the judgmental elements are not obvious, the users of models--especially complex models--can easily confuse the model and reality. They come to believe they have knowledge about the energy system when they have knowledge only of the model. And in the saying quoted by Freedman, Rothenberg, and Sutch (1983b): "It ain't what you don't know that gets you into trouble, it's what you think you know that ain't so."

When policy analysts equate a model with reality, they begin to define issues in the terms most central to the model and to ignore the variables the model ignores. Among these ignored variables are many important behavioral factors for which quantitative data do not exist and that are not prominent in economic theory, for example: incompleteness of information, mistrust, the symbolic meanings of action, personal attitudes, social values, political conflict, and organizational routine. The phenomena that result from such influences are, of course, encompassed by formal demand models, but they appear under other labels and so may be misconceived in important ways.

The Place of Behavioral Variables in Formal Demand Models

In energy demand models, behavioral variables are usually subsumed under other broad concepts that are vague with respect to their behavioral basis. The concept of own-price elasticity, for example, describes the fact that, other things being equal, the quantity demanded of something is inversely related to its price. But the concept says nothing about how the information embodied in price enters a consumer's awareness or about how awareness of price affects action. Thus, the concept of elasticity bypasses the behavioral phenomena that underlie the response to price; the behavioral questions are begged. Because the concept of elasticity is behaviorally atheoretical, elasticity estimates cannot predict whether unanticipated conditions, such as news of an impending major oil shortage, will change the ways consumers respond to price signals. A focus on price elasticity also ignores factors that can mediate the effects of price,

such as the quality or trustworthiness of available information or the ways that information is communicated.
Because behavioral research focuses on the underlying
processes, it may help demand analysts make forecasts
about new conditions. In this section, we discuss two
concepts central to many formal demand models--the discount rate and the dynamics of response--to illustrate the
ways behavioral concepts relate to the concepts used in
formal models.

The discount rate is a key concept in many demand
models. Formal models estimate the ultimate penetration
of energy-efficient technologies by making the assumption
that energy users will, in the long run, make the economically rational choice of the alternative with the greatest
net present value in terms of investment cost plus subsequent operating cost. What is economically rational for
a consumer is a function of that consumer's discount rate,
that is, the magnitude of the consumer's preference for
present over future value. Thus, the concept of discount
rate implies using a particular mathematical formula to
account for the fact that the real dollar value of investment in energy efficiency is generally less than the real
dollar value (undiscounted) of the expected return.

The magnitude of the discount rate is often simply
postulated in formal models, but some empirical methods
have been suggested for estimating discount rates. The
methods usually used involve analysis of data on consumer
choices among alternatives that vary in energy efficiency
and investment cost (e.g., for appliance purchases,
Hausman, 1979). From those choices, an implicit discount
rate is calculated: it is defined as the discount rate
that would make actual purchase behavior economically
rational in terms of net (i.e., discounted) present value.

The use of discount rates in models presents a major
practical problem in that data are often inadequate for
confidently estimating implicit discount rates. Sometimes
markets do not offer a wide choice of energy efficiencies
or do not present a trade-off between energy efficiency
and initial cost, and so it is difficult, even in principle, to collect the relevant data. In other cases, as
with some household appliances and home insulation, the
needed information could be obtained through additional
efforts at data collection and analysis; however, great
effort might be required because implicit discount rates
probably vary with characteristics of the investment and
the consumer (e.g., for different income levels, Hausman,
1979; McFadden and Dubin, 1982) in ways that cannot

readily be predicted from the concept of time discounting
alone. But the analyses are necessary if formal modeling
is to be based on reliable estimates of discount rates.

The use of discount rates can also breed a broader
conceptual problem. The concept of time-discounting
assumes that consumers make decisions based on the value
they place upon time and that the relevant decision rule
can be represented by a number that is constant for particular types of consumers, or at least for individual
consumers making particular types of purchases. Because
time-discounting is implicitly a psychological process of
the consumer, it is easy to interpret empirically calculated implicit discount rates as representing an attribute
of energy users.[5] But when implicit discount rates are
calculated from data on purchases, the resulting number
reflects a collection of variables--not only the degree
of preference for present value--that may affect the
level of investment. The calculated discount rate is
affected, for example, by the extent to which information
is imperfect, mistrusted, or ignored; by energy users'
persistence in old habits; by the fact that consumers
with limited capital do not always purchase what they
would if they had more capital; and by various other
factors that might change the rate of adoption of
energy-efficient technology.

To subsume all these influences under a single index
and to call it the discount rate may be to misconceive the
phenomena. More important for forecasting and policy
analysis, such theoretical shorthand may lead analysts to
think of some features of energy users' behavior as stable
preference (part of the discount rate) when they may in
fact be changed by economic or institutional forces or by

---

[5]Of course, the argument can be made that consumers act
as if they were making decisions about the value of time
without going through any such psychological process. In
this view, the discount rate is simply a mathematical
shorthand for projecting the difference between the real
dollar value of investment and the expected future return.
This interpretation eliminates any psychological rationale
for the functional form of net-present-value equations.
It also lacks any theoretical structure that might offer
insight into the variables that determine discount rates
or that might suggest policies for changing the rate of
acceptance of new technology.

policy. When important variables that influence behavior
are treated as constants in analysis, erroneous forecasts
may result. Analysis in terms of discount rates may also
lead analysts to overlook potentially effective policies
because the effects would be on something that has been
implicitly defined as a stable attribute of energy
consumers.

Another central concept in many demand models is that
of response dynamics. Investments in energy efficiency
in response to new conditions occur slowly over time.
Formal models commonly estimate the rate of investment a
priori, using an assumption that the rate follows price
changes or other stimuli with a time lag described by a
particular mathematical function. A typical example is
the original Oak Ridge National Laboratory model of residential energy use (Hirst and Carney, 1978), which used a
simple algorithm to estimate the extent and pace at which
manufacturers and households improve the energy efficiency
of equipment and structures in response to changes in
fuel price. Neither theory nor data were available to
validate the assumptions embodied in the algorithm, and
little work has been done since then to correct the
deficiency. There is some evidence, however, that the
most typical algorithms do not give the best estimate of
the dynamics of response to changing energy prices (Hill,
1983). Thus, formal models lack an empirical basis for
estimating the dynamics of price response.

Another aspect of the problem is the way time-lag
specifications are used in econometric models. Lag
coefficients quantify a phenomenon without identifying
any of its multiple causes.[6] The slow response to new
conditions is sometimes explained in terms of budget constraints or the costs of replacing capital stock before
the end of its useful life, but there are many other possible explanations. To cite two examples, full informa-

---

[6]Modelers sometimes estimate dynamics from assumptions
about decision processes rather than from assumptions of
correspondence with a mathematical equation. Since this
procedure is based on implicit or explicit theory, it may
have testable implications about the causes of dynamics.
But because the data needed to support the assumptions do
not exist, postulating decision processes is not much of
an improvement over postulating the functional form of a
descriptive equation.

tion spreads slowly, and new information sources take time
to become credible. No doubt, many other behavioral and
institutional factors also act to retard the adoption of
energy-efficient technology. As with the discount rate,
it would be a mistake to interpret a lag coefficient in
terms of only one of the factors that affect it. Thus,
for forecasting or answering analytical and policy questions, a model of energy demand needs more than a lag
coefficient that accurately describes past events. The
coefficient should be based on an understanding of the
ways various environmental factors, including behavioral
ones, affect rates of response.[7]

Formal demand models tend to address behavior through
concepts that cover the outcomes of behavioral phenomena
but offer little or no insight into the determinants of
the phenomena. This approach is useful for forecasting
only as long as the relationships among the important
behavioral variables are relatively stable--but there is
little reason for assuming such stability given the
unprecedented changes that have been occurring in the
energy environment. For policy analysis, the approach has
the limitation that it directs attention away from some
behavioral variables that may offer useful levers for
policy.

There are two possible ways to resolve these problems,
and they are not mutually exclusive. One way is to direct
resources to gathering the data needed to test key assumptions and provide empirical grounding for the parameters
used in policy models. Many data are needed; we discuss
their specifics in the following chapters and offer recommendations in Chapter 6. The other way is to direct
resources to other approaches to policy analysis that
focus on understanding the behavioral phenomena of interest in policy contexts. This problem-oriented way of
doing energy demand analysis is discussed in the next
section.

---

[7]Models relying on lag coefficients can be effective in
making projections under some conditions. But without
understanding of the factors responsible for the dynamics
of response, it is impossible to know what changes in the
environment would invalidate the projections or what
policy alternatives might alter the dynamics from past
patterns.

PROBLEM-ORIENTED ANALYSES OF ENERGY POLICY ISSUES

In contrast to formal policy models, problem-oriented analyses are by definition unsystematic, being addressed to specific questions rather than to the accurate description of an entire social or economic system. An issue or policy option is raised, and information is gathered and analyzed by the best available methods to clarify the issue or the policy implications. This section describes the major research methods used; we discuss their uses in more detail in later chapters.

Methods of Problem-Oriented Analysis

Surveys

National general-purpose surveys of energy users can collect data simultaneously on a range of variables relevant to energy use. Such surveys can allow for multivariate analysis of the relationships among economic, demographic, climatic, engineering, and social-psychological variables as they affect energy demand. They can include variables not presently addressed in models and can estimate their importance. Such surveys would use a nationally representative sample of energy users. If repeated, surveys can gather time-series data essential for empirically estimating formal demand models, particularly econometric models.[8]

National surveys are not, however, useful for all types of policy analysis. General surveys have little value, for example, for analyzing policies or programs that have not yet been tried or that few people have experienced, such as new appliance labels, utility load management programs, or comprehensive home retrofit programs, because few people in a general population sample will have participated in such programs. This problem can be addressed by conducting more specialized surveys on samples drawn from program participants and comparison groups.

---

[8]Researchers distinguish among cross-sectional survey designs, which collect data only once; longitudinal designs, which collect data on the same variables over time; and panel designs, which collect data on the same variables from the same set of respondents over time.

Specialized surveys--surveys aimed at particular issues or types of energy users--can help answer important analytic questions that cannot readily be addressed by models or in general national surveys. Thus, a national survey focused on solar energy can intensively probe for the factors determining intentions to invest in residential solar energy technologies (Farhar-Pilgrim and Unseld, 1982), or a utility company can survey its customers to assess the reasons for their response or nonresponse to a conservation program (see Berry, 1981). Specialized surveys are valuable because they can look closely at variables, including many of those not included explicitly in formal policy models, that may affect important consumer actions. For example, a survey in Michigan estimated the magnitude of consumers' misunderstandings of household energy use (Kempton, Harris, Keith, and Weihl, 1982).

But surveys suffer from some generic limitations. Surveys may have poorly worded questions, may induce respondents to give socially acceptable rather than accurate responses, or may fail to predict behavior because the respondents themselves cannot predict what they will do. Unreliability increases when surveys are used to assess responses to hypothetical situations, such as an experimental electricity rate structure; people cannot reliably predict their behavior in situations they have not experienced. A further problem with using survey results to project energy demand is that even when self-reports are accurate, inferences from them may not be. A major reason for this lack of correspondence is that energy-saving actions, such as investment in home insulation, do not always save the amount of energy that would be expected from technical estimates. This "failure" may be because energy use data are unreliable or because technical estimates make unrealistic assumptions about how well the improvements are installed; it is also possible that people use some of the energy saved by technical improvements to increase their comfort (Hirst and Goeltz, 1983; Hirst, Hu, Taylor, Thayer, and Groeneman, 1983; Hirst, White, and Goeltz, 1983a). Thus, the best way to gauge the likely response to a new information or incentive program is to rely on actual observation.

Specialized surveys often have problems attributable to their low budgets. Their sample sizes are small, the research staffs are often inexperienced in the finer points of survey design, and their findings are often poorly documented. There has frequently been a wide gap between intent and execution in specialized energy surveys.

Analysis of Existing Data

The Energy Information Administration has several data
sets that have only been partly analyzed, and utility
companies have some of the best existing data on
residential and commercial energy use. Utility data are
particularly useful for analyzing responses to price
because in some utility service areas, rapid price
changes have recently occurred for one fuel supplied but
not another, and within the regions, electricity prices
have changed rapidly for some utilities but not for
others.[9]
   Analyses of past experience may fail for lack of critical information, or they may be misleading for making
projections because of differences between past and future
situations. There have been problems getting access to
existing data at the individual level because of concern
about privacy. And there are also obviously limitations
on the kinds and quality of the data available. Better
data exist for analyzing energy use in the residential
sector than in the commercial or industrial sectors;
aggregate data are generally more available than disaggregated data; energy-use data are better than data on
equipment stocks; and data on attitudinal factors is particularly adequate. In addition, some data, such as on
demographic variables and local weather conditions, are
often not available in a merged form with disaggregate
data on energy use.

Natural Experiments

Rapid changes in the prices of fuels over the past decade
have made it possible to study the price effects empirically, and the fact that price changes were not uniform
across fuels or localities provides useful comparison conditions. The recent sudden decrease in the inflation rate
constitutes a natural experiment on the determinants of
consumers' price expectations. Such natural experiments

---

[9]Existing data can be analyzed in various ways. One of
these involves using the same econometric techniques that
are sometimes used for building detailed formal models.
Thus, small-scale, problem-oriented modeling can be part
of an analytic alternative to system-wide formal demand
modeling.

could provide much of the data needed for demand analysis with relatively little additional effort, if the data are collected regularly. Methods for evaluating the results of such natural experiments have been developed over the past decades (e.g., Campbell and Stanley, 1966; Cook and Campbell, 1979), but have rarely been used to learn from natural experiments on energy demand. Of course, a natural experiment is riskier to interpret than a controlled experiment because it is harder to identify the causes of the phenomena observed.

Controlled Experiments

The experimental approach has been generally neglected in energy policy analysis. The best-known exception was the time-of-use pricing experiments conducted during the 1970s, some of which involved random assignment of households to experimental electricity rates. Experimentation was the method of choice in those studies because there was no empirical basis for estimating the effect of prices based on time of use and because the experimental rates were so far from most energy users' past experience that self-reported intentions could not be relied upon. The same rationale suggests that experiments could provide the most valid answers to questions about the design of energy conservation programs, particularly for assessing the effects of interventions that are nonfinancial in character and for which, as a result, existing models are particularly inadequate.

Even laboratory experiments are sometimes appropriate for policy analysis. As part of energy information efforts, it is necessary to choose what information to provide, how to design the layout of appliance labels, and so forth. For many of these choices, it would be useful to experiment with alternatives in a laboratory setting to see which alternatives are eye-catching, understandable, meaningful, and considered useful by people like those for whom the information is intended.

The greatest advantage of controlled experiments over other research techniques, of course, is that they can control for large numbers of extraneous variables that may covary over time with the variables under study and make the interpretation of nonexperimental data difficult. For this reason, experimentation is sometimes useful even when models, analyses of existing data, or survey research can produce empirically meaningful results. Thus, small-scale field experiments on the effects of rebates as an incentive for energy conservation (Geller, Winett, and Everett, 1982) can provide a check on the elasticity estimates derived from analysis of time-series data.

Like other research methods, experiments have their problems as a policy tool. Some researchers, unfamiliar with practical policy concerns, have experimented with unrealistic treatments and consequently produced impractical recommendations (see Stern and Oskamp, 1984). Experimental studies often meet practical opposition from program managers who are eager to get on with their programs and who feel they know enough to act without awaiting the results of formal research. Experiments also face political opposition as unethical or unnecessary: if the policy is a good one, it should be made available to all, not just a small experimental group (for a discussion of such issues, see Mosteller and Mosteller, 1979). Moreover, if experimental subjects believe an experiment to be temporary rather than a permanent change in policy, their belief may affect their behavior.

Evaluation Research

Past and present energy programs are a great untapped source of information about energy demand. Information programs run or mandated by government--such as Project Conserve, the Residential Conservation Service, the Energy Extension Service, and the energy-efficiency labeling of cars and appliances--all constitute experiments with information, but they are almost always uncontrolled, and few have been adequately evaluated. Incentive programs-- such as the federal and state tax credits for conservation and solar energy--have also been inadequately studied. And thousands of local energy programs, public and private (see Center for Renewable Resources, 1980) have been started, but rarely with an evaluation component (Stern and Aronson, 1984:Chapter 7). Evaluations of such programs can provide knowledge that is unlikely to result from other research methods. In particular, evaluation studies can uncover variables in the implementation of conservation programs that would not be anticipated by before-the-fact analyses.

Evaluation research can use any of the methodologies outlined above. The firmest conclusions, of course, can be drawn if energy programs are treated as experiments from the start. A variety of quasi-experimental research designs that retain many of the advantages of experimentation can also be used. The call for experimental control does not, however, imply that energy programs should be rigidly specified in advance for scientific purposes; that approach would limit the programs' abilities to adapt to their environments and could threaten their success. Rather, we are emphasizing the advantages of experimentation that accrue from procedures of careful measurement and of random assignment to treatment conditions, which control for a variety of extraneous variables.

Evaluation research usually has serious inference problems. Because the call for evaluation usually comes after a program is in place, experimental design is impossible, and researchers sometimes attribute an effect to a program that may have been due to self-selection of program participants. Furthermore, when a study begins after a program, some of the essential data may not be available, and preexisting conditions cannot be reliably assessed.

The Strengths and Weaknesses
of Problem-Oriented Research

Individually, the above research methods are limited in their usefulness for analyzing energy demand; together, they constitute a valuable set of tools. They provide policy analysts with several ways to answer particular policy questions. For example, to assess response to a tax credit for investments in energy efficiency, several approaches could be combined: surveys of willingness to invest given the proposed tax credit; an evaluation study of the effects of a similar incentive that has been tried in another state or country; small-scale field experiments in which the tax credit or a similar incentive was actually offered and actual investment by participants was assessed. Analysis based on several methods is less prone to errors that arise from reliance on a single method.

Problem-oriented studies can also be used to address issues that arise in constructing policy models. When data are not available for estimating discount rates, surveys to assess consumer choices among hypothetical technologies may, despite the limitations of such surveys, be the best available source of estimates. To estimate price elasticity over a range of prices that has never been experienced (e.g., premium prices for peak-period electricity), small-scale experiments are advisable and have sometimes been conducted.

Problem-oriented research is also useful for answering behavioral questions that are not explicitly addressed in models. If a policy maker needs to know how to design an energy-efficiency label for an appliance so that consumers understand it, or how much difference a well-designed label can make in consumer response, or whether consumers are likely to believe the information the government requires in a label, formal models are not equipped to offer answers. But laboratory experiments can help answer the first question, a field experiment can help answer the second, and surveys and field experiments can help answer the third.

Compared with modeling, problem-oriented research offers certain advantages. It can often address questions more directly than models can, and it can investigate a number of questions that cannot readily be represented in models or for which representation is possible but no empirical data exist. Before modelers respond to a new policy question with an empirically unsubstantiated ad hoc adjustment to an existing model, a problem-oriented study can provide the data needed to address the question empirically. In addition, some methods of problem-oriented research, particularly experimental ones, provide more convincing evidence than modeling data can offer. Experiments, and even well-conceived quasi-experiments, are relatively free of the problems of spurious correlation that haunt econometric models and of the unsubstantiated causal assumptions that leave system dynamics models open to serious question.

The problem-oriented approach has conspicuous weaknesses. It is not by itself well suited to forecasting energy demand. It can easily overlook the possibility of complex or counterintuitive interactions of a proposed policy with other events in the economy or energy system. Because of its responsiveness to changing policy concerns, the approach may not benefit from systematic accumulation of knowledge. And if problem-oriented studies proceed from a lack of basic understanding of the behavioral variables policy is intended to influence or of the practicalities of policy implementation, they will tend to frame questions in unproductive ways and yield impractical advice.

In addition, each research method used for problem-oriented analysis has its own particular weaknesses, as noted above. An extensive literature on social science methodology details the strengths and weaknesses of the various research designs that can be used for analyses of

questions relating to energy demand (the classic brief
statement is by Campbell and Stanley, 1966; see also Cook
and Campbell, 1979).

As with formal policy models, many of the weaknesses
of individual methods of problem-oriented research are
surmountable. Replication and criticism are two obvious
methods of quality control. In addition, each research
method provides a way to cross-validate results obtained
by other methods. Since findings are best established if
they are proved robust to choices of methodology, problem-
oriented energy research may make its strongest contribu-
tion as a method of quality control and validation.

Although problem-oriented research cannot offer the
broad view that energy models attempt to provide, it can
be an invaluable part of the effort to understand the U.S.
energy system. It offers ways to validate assumptions,
to estimate the parameters of models, to see if variables
that have not yet been considered may be important, and
to explore in detail the behavioral phenomena that underlie such broad concepts as price elasticity, time lags,
and implicit discount rate. Problem-oriented research on
energy demand continues to occur in response to the needs
of policy makers and the interests of researchers. But
the insights from this research have not been systematically incorporated in formal analyses of the energy system, and the informational needs of formal modelers have
not yet served as a significant impetus for problem-
oriented research.

CONCLUSIONS

There are many critical gaps in knowledge about energy
demand. Given the present state of knowledge, it is clear
that most analyses are based on at least some erroneous
assumptions and ignore at least some important variables.
Formal demand models, in particular, are unreliable guides
to policy analysis. Too often, there is no evidence to
confirm or revise the assumptions in such models, to
decide if the important variables have been correctly
identified, to support parameter estimates, and to evaluate the importance of variables omitted from the models.
Not only is convincing evidence lacking to justify the
assumptions, but data essential for obtaining the evidence
are also lacking. This state of knowledge is an important
reason that demand models are the subject of so much
debate.

Problem-oriented studies have great potential value for filling gaps in demand analysis, but--even within their limits as analytic tools--they have been insufficiently used. Problem-oriented research can produce significant new knowledge, but we believe this knowledge will not always be readily translatable into the language of existing models. Rather, we expect that improved knowledge will bring about changes in existing models and that such changes will lead to improved analysis of energy demand. This report is a beginning effort to show how different analytic methods can be used in a complementary fashion to improve understanding of energy demand.

2

The Effects of Price on Demand

There is no doubt that price affects demand, but the way
that happens is not yet clearly established. Behavioral research gives reason to question some of the most
common assumptions about what features of price consumers
respond to. The research suggests that price may be a
more highly differentiated variable than the one estimated
in most formal demand models.

This chapter discusses five different dimensions of
price stimulus that can plausibly have important effects
on individual demand for energy at any given time: (1)
average price versus marginal price; (2) real price versus
nominal price; (3) price levels versus price changes; (4)
price-change thresholds versus linear price effects; and
(5) price increases versus price decreases. For the first
three dimensions, we examine the empirical evidence, discuss its practical significance, and consider possibilities for future research. The issue of the formation of
price expectation is covered briefly in the context of
responses to price change, but the question of how consumers form long-run expectations when making long-run
decisions is not covered. The last two dimensions of
price are discussed more briefly.

Our ability to explore these dimensions of price
reflects the types of data available. Prior to the early
1950s, almost all econometric analyses were confined to
examining the changes of highly aggregated economic measures (e.g., national gasoline consumption and average
retail price) over time. Since many policies are concerned with such aggregates, such time-series analyses are
potentially useful for forecasting. However, the usefulness of aggregate data in understanding individual behavior is very limited. Since the 1950s, data at the individual or household level of analysis have become much

more readily available. These data have usually been cross-sectional, that is, based on measurements at one time. One can hope to infer the causes of individual behavior with such data by comparing individuals presumed to be similar in all respects except for the level of the causal variables in question. Inference is complicated by the fact that not all factors that affect outcomes for individuals can be identified or measured. If there is any association of unmeasured factors with the variables of interest, the estimates of those variables' effects may be biased. Similar problems occur when unknown functional relationships are misspecified. With the time-of-use electricity pricing experiments of the 1970s, some experimental data have become available on consumer demand behavior. However, because of the flaws in the designs of many of those experiments (see Hill et al., 1978 for an evaluation of the designs), very few analyses of these data have estimated treatment effects.

Most recently, longitudinal data sets have been developed that use households as the unit of measurement and that follow panels of households over time. With these data it is possible to measure change in outcome variables directly and to relate those changes to the changes in antecedent variables. While the resulting estimates of the effects may still be biased, this bias may be reduced because the repeated measurement of the same households holds constant the effects of any unmeasured individual differences that are stable over time. For the purposes of this chapter, then, most of the useful evidence is from time-series data; only one or two of these studies, however, are longitudinal.

DO CONSUMERS REACT TO AVERAGE OR MARGINAL PRICES?

Although elementary economic analysis may lead one to believe that consumers respond to marginal prices, there is some behavioral evidence that householders, at least, do not seem to notice them. Kempton and Montgomery (1982), for example, conducted detailed interviews with Michigan householders and concluded that the most common units people used to quantify their energy use was dollars per month--a measure of average price. If such awareness shapes behavior, householders will not respond as might be expected to the block-rate schemes common in utility billing; changing the price differential between blocks, for example, would not affect the demand of users whose

total bills remained unchanged; neither would inverting
the rate structures. When a price change is restricted
to the marginal block of energy used, the demand of consumers who response to average price will be less elastic
than if they responded to marginal price; and base price
increases, such as for service, might decrease energy use
even though they have no effect on marginal prices.

Of the questions discussed in this chapter, the question of whether consumers react to average or marginal
prices is the most thoroughly researched by economists.
Most formulations of demand theory assume that consumers
are able to purchase various goods at a constant per-unit
price. When this is true, average and marginal prices are
identical. But because there are large fixed components
to the costs of supplying energy, marginal production
costs are generally lower than average costs. As a
result, some (often complicated) price schedules have
evolved.

The declining block-rate schedule has, until recently,
been the predominant form of pricing electricity and natural gas in the United States.[1] The price per unit for
consumption of these energy goods is a declining function
of the level of consumption (usually calculated on a
monthly basis). For a set minimum level of consumption,
customers are charged a constant per-unit rate. If a
customer consumes more than the minimum, the additional
amount is billed at a lower marginal rate, which is in
effect for consumption between the minimum and a set
higher level of consumption. Beyond that level, a still
lower rate is charged. While most utilities only distinguish two or three such rate blocks, some have more. In
any event, such a pricing schedule results in differences
between marginal and average prices, and those differences
can be quite large.

When marginal and average prices are different, the
question arises as to which is more important in determining consumer behavior. The answer that can be derived
from demand theory is both; theory also predicts that
consumer demand will also be affected by the structure of

---

[1]Over the last decade many utilities have instituted
"life-line" rates and other forms of inverted block
pricing schedules. The fact that new electric generating
capacity now costs more than the average cost of electricity has contributed to this change.

the entire rate schedule.[2] Economists explain this
through an analysis of the declining block-rate structure.
They define the marginal price as that charged in the
consumption block corresponding to the highest levels of
energy use; the rest of the energy cost--consumption in
lower blocks times the price differential between those
blocks and the highest block--is called a tax. This "tax"
is sometimes called an "income component" (see, e.g.,
Taylor, 1978; Taylor, Blattenberger, and Verleger, 1980;
Nordin, 1976) because its effect is to decrease consumers'
real incomes by an amount that may be affected little or
not at all by marginal energy use.[3] The theory is used
to argue that changes in average prices when the marginal
price is constant have an impact equal in magnitude to the
effect on demand of an ordinary change in income. This
analysis implies that demand is a function of both marginal price and of the "tax" on lower blocks of consumption, which affects energy use through the budget con-
straint. The size of the "tax" is a function both of
average price and of the starting point of each consumption block. In theory, this combination of factors deter-
mines the shape and position of the demand function and
therefore the equilibrium level of demand for energy
priced under a given rate schedule.

Empirical evidence also suggests that both average and
marginal price play a role in determining behavior.
Invariably, however, the effect of average price has been
found to be larger than that of an ordinary change in
income--in one case (Hill et al., 1978), by a full order
of magnitude. The data suggest, then, that consumers
respond to average price as distinct from marginal price
and also that they are more responsive to average prices
than to changes in income. However, considerable care
must be exercised in interpreting this consistent finding
from the various models. While the per-dollar effect of
changed cost in lower consumption blocks is, in most

---

[2]Indeed, Blattenburger (1977) has shown that, in theory,
some changes in consumption can be induced by changing
rate schedules, even though average and marginal prices
remain unchanged.
[3]This "income component" is also referred to in the economics literature as the "rate schedule premium" (Barnes,
Gillingham, and Hagemann, 1982), "intra-marginal expenditures" (Taylor, 1978), and "implicit income" (Hill, Ott,
Taylor, and Walker, 1983).

studies, much larger than that of normal income, the cost changes ranged from $2 to $20 per month, while incomes ranged from $200 to $4,000 or more per month. Thus, regular income is a far more important determinant of energy consumption. Similarly, while it is generally found that costs in lower blocks (as measured by standardized regression coefficients, t-ratios, marginal R^2s, and so forth) are more strongly predictive of demand than is marginal price, it must be recalled that these measures are designed to capture all of the "income effect" of the rate schedule and that they are definitionally related to consumption. Marginal prices are, in the common analytic construction, not definitionally related to consumption, the entire motivation of the exercise generally being to isolate the true impact of marginal prices.

Even with these caveats in mind, it seems safe to conclude that consumers react more vigorously to rate-schedule components associated with average prices than economic theory suggests. That is, consumers seem to react both to average and marginal price, although it is not yet possible to estimate the relative magnitude of these effects.

What practical difference is the effect of average price likely to make, and what can be done to improve policy analysis? Even though cost changes in lower consumption blocks are small and their effects on energy consumption are likely also to be small in comparison with income effects, consumer reaction to average price may have potentially important ramifications for policy models. Rate schedules are currently undergoing major restructuring: declining block-rate schedules are rapidly being replaced by life-line rates, "inverted" or increasing block rates, and even time-of-day seasonally adjusted flat rates. The divergence of average from marginal prices that may arise from these new structures may be much greater that that observed in the past for declining block rates, and responses to average prices may gain correspondingly in analytical importance.

There are ways to assess this possibility, and the empirical literature on the effects of differences between average and marginal prices contains an important key. When the issue of responses to average price first arose, economists were forced to base their estimates on the average electricity prices average consumers paid. With considerable effort, analysts progressed to using industry-compiled information for "typical electric bills" at 100 kWh/month, 250 kWh/month, and 750 kWh/month for consumers in a given state. Finally, with even more effort, some researchers have recently used information about the actual rate structure individual consumers face to explain their behavior. Understanding has grown in proportion to the level of detail and disaggregation of the data available for analysis.

The fact that rate structures are presently changing rapidly means that an important natural experiment is taking place. Energy analysts are not monitoring it in any systematic fashion. Following even a small number of energy consumers through the course of this natural experiment could be one of the best possible investments in energy research.

DO CONSUMERS REACT TO REAL OR NOMINAL PRICES?

By the definition of economically rational choice, that choice is a function of real price. But according to self-reports, residential energy users mainly pay attention to nominal prices (Kempton and Montgomery, 1982). This possibility, that people suffer what economists call a money illusion, is consistent with a body of literature in cognitive psychology that shows that people use cognitive heuristics to simplify complex choices for themselves (Kahneman, Slovic, and Tversky, 1982) and that people's choices sometimes depend not only on the expected values of the alternatives but on the way the alternatives are described (e.g., Tversky and Kahneman, 1981). Thus, it is plausible that in times of inflation people may operate on a rule of thumb that "all prices are going up," and thus fail to respond to a real price increase for a particular commodity. Or a decline in real income may be taken as a signal to cut all expenditures, even for items whose real prices are declining faster than income.

Economists have explored such possibilities under the rubric of the homogeneity of demand functions. If consumers base their expenditures on real prices, proportionally equal increases or decreases in all nominal prices and incomes should leave the demand for all goods unaltered. Relationships that exhibit this property are termed "homogeneous of degree zero." When demand functions are homogeneous, the demands for all commodities add up to total income. The exact form of this restriction depends on how preferences, and thereby the demand functions, are specified. To test the implications of homogeneity, it is necessary to analyze the demand for all

goods at once. Although this is not feasible for all
goods taken individually, it has been accomplished in
some models for broad classes of goods (such as food,
housing, entertainment, and so forth) as well as for some
subsets of goods, such as peak and off-peak electricity.

Almost without exception, in studies of demand for
meat, consumer durables, and electricity, the restrictions
on behavior implied by homogeneity have been rejected
(see., e.g., Atkinson, 1977; Byron, 1970; Caves and
Christensen, 1978; Court, 1967; Diewert, 1974; Theil,
1965). However, in all of the studies the hypothesis
rejected was actually a joint hypothesis: that demands
are homogeneous and that the particular specification of
the demand function used is appropriate. The rejection
of this joint hypothesis does not allow one to determine
which part is incorrect.

There is some evidence that it is the specifications--
which are usually highly simplified--that are incorrect.
The residential time-of-day electricity demand systems
used by such authors as Atkinson (1977) or Caves and
Christensen (1978) provide a good example. These models
use single-valued demand functions (i.e., functions allowing only one level of demand for each price) and ignore
the fact that intervening between electricity consumption
at various hours of the day and consumers' utility is a
complicated set of home production relationships. If the
production of commodities at home were a smooth and continuous function of peak and off-peak electricity usage
and if the law of diminishing returns also applied, there
would be no problem. But some home-produced commodities,
such as hot water, are produced primarily in batches and
stored from one period to the next, to take advantage of
economies of scale; householders consume from the resulting inventories in accordance with classic inventory
adjustment processes (Hill, 1978). This means that for
some uses electricity in one period is substitutable for
that in another and, therefore, that for certain combinations of peak and off-peak prices the derived demand
functions are multivalued, nondifferentiable, and even
discontinuous. While some possible aggregations of
activities might tend to smooth out the production sur-
face, it remains true that home production is complex and
its outputs may not be representable by a smooth function
of peak and off-peak electricity demand. Thus, it would
be fortuitous if a single-valued continuous demand specification would yield results consistent with homogeneity
(or any condition implied by demand theory except negativ-

ity), even if preferences were perfectly well-behaved when defined in terms of home-produced commodities such as clean clothes and cooked meals.

Further indirect evidence that it is the specification of demand functions that is in error rather than the assumption of homogeneity is provided by nonparametric tests based on the axioms of revealed preference.[4] In general, the restrictions of demand theory (for example, negatively sloped demand functions) are found by those tests to hold for aggregate data. Unfortunately, however, the homogeneity hypothesis is not directly testable with these axioms (see Deaton, 1983). In conclusion, the empirical evidence contained in the literature on demand systems does not rule out the possibility that consumers react to nominal rather than to real prices.

Two other sources of evidence in economics may be pertinent to the question of responses to nominal price: macroeconomics and the study of equilibrium in single markets. In the macroeconomic literature, the issue of real versus nominal prices is generally cast in terms of differential perception of price and income changes. If consumers perceive only price increases, inflation should increase saving because the price increases would dampen consumption; the reverse type of money illusion (i.e., consumers' perceiving only income increases) would reduce saving.[5] The evidence has suggested that money illusion is a relatively short-term phenomenon. Until quite recently, savings rates had remained fairly constant for 200 years while price levels had increased manyfold. However, differential perceptions of and responses to price and income changes may be important in determining individual behavior in the short run.

Evidence from the study of single markets (e.g., studies of gasoline demand) is hard to consolidate because there are so many markets. Most of the studies use only real (i.e., deflated) prices and therefore do not inves-

---

[4]The axioms of revealed preference were originally developed by Samuelson (1970) and have been extended and operationally defined by Afriat (1973), Diewert (1973), and Varian (1982). Much empirical evidence in the psychological literature, however, raises serious questions about the adequacy of the axioms (e.g., Tversky and Kahneman, 1981; Kahneman, Slovic, and Tversky, 1982).
[5]However, if consumers expect continuing (or escalating) inflation, different behavior would be expected.

tigate the question of whether it is real or nominal
prices that best explain behavior. When comparisons are
attempted, they are hampered by the fact that nominal
prices for particular commodities and general price
indices tend to covary. In cross-sectional analyses,
there is only one value for a commodity price and one for
the general price index, and so it is impossible to examine the issue. Dividing the variable nominal prices by
the constant price inflator may change the nonstandardized
price coefficient (because it changes the scale of measurement), but it will not alter any goodness-of-fit measure of the estimating algorithm and therefore provides
no means of judging which form of price best explains the
data. With longitudinal or time-series data, as well as
with cross-sectional data containing local price-level
information, the problems in analyzing the data are less
statistical and more conceptual. If the observation
period is long enough and varied enough, there may be
sufficient independent variation in nominal prices and
price-level indices for statistical methods to distinguish
which price concept reveals more about behavior, but
choosing the weights upon which the index is based
requires care and can dominate the results.

Thus, the empirical evidence on whether consumers
respond to real or nominal prices is inconclusive. However, economists seem to agree that while the speed of
reaction may well depend more on nominal prices, the overall reaction is constrained by real prices. The importance of the issue of consumer reaction to real versus
nominal prices depends on the level and stability of the
overall rate of price change. When inflation is low and
constant, as in 1955-1965, it does not really matter if
consumers are reacting to nominal or real changes since
the two are nearly equal and move in the same direction.
During periods of higher inflation and, especially, variable inflation, the issue may be more important. The last
10 years have been such a period. Recent price instabil-
ity offered a good opportunity to collect relevant data
on price response, but energy analysts did not do so.
There may be other chances in the future, and analysts
should be ready to take advantage of them.

DO CONSUMERS REACT TO PRICE LEVELS OR PRICE CHANGES?

Although demand analysts usually express equilibrium conditions in terms of price levels, economists have hypothe-

sized for years that the speed of consumer adjustment to new prices is proportional to the rate of change of prices. Such a hypothesis is consistent with an extensive empirical literature in the psychology of perception (e.g., Helson, 1964). The most common form of the hypothesis in economics is the adaptive expectations model in which consumers are assumed to allocate expenditures on the basis of "expected" long-term (or normal) prices rather than actual current prices. Expectations are assumed to be adapted by a constant fraction of the discrepancy between most recent expected and current actual prices. That is,

$$P_t^* - P_{t-1}^* = (1 - r)(P_t - P_{t-1}^*) \ ,$$

where P_t^* is the price expected in period t for period t + 1; P_{t-1}^* is the price expected in period t - 1 for period t; P_t is the actual price in period t; and r is a positive number less than one. If prices have been stable until the current period, it is easy to show that the current expected price equals the last period's actual price plus a fraction of the change in actual price between the two periods: that is,

$$P_t^* = P_{t-1} + (1 - r)(P_t - P_{t-1}) \text{ when } P_{t-1} = P_{t-1}^* \ .$$

Thus, if consumers do base their decisions on expected prices and if they base their expectations on deviations of actual from expected prices, then both the level and the rate of change of prices should affect consumer reaction.

The adaptive expectations model has been used numerous times in energy demand studies (see, e.g., Dahl, 1979; Houthakker, Verleger, and Sheehan, 1974; Kwast, 1980; Taylor, 1978). The data are almost always found to be consistent with the hypothesized model. The trouble is that in most cases, the hypothesis is not incorporated into the model directly but via the Koyck transformation (Koyck, 1954). This technique makes it nearly impossible to distinguish whether the data are being generated by adaptive expectations, by partial adjustment to current prices, or even by simple serial correlation of the errors (see Griliches, 1967) because all three hypotheses imply the same things for the estimated coefficients in the model.[6]

---

[6] In theory one should be able to tell which mechanism is operating because the adaptive expectations model

The work of Archibald and Gillingham (1978a, 1978b) is notable in that it allows alternative hypotheses to be distinguished. In their longitudinal study of 1972-1973 consumer expenditure data, they include both current prices and quarterly rates of change in price in a regression model of gasoline demand. They find that demand decreases as price levels increase but, holding price levels constant, is higher when the price increases have been more rapid. This model fits the data better than a related alternative that includes price level and its square. While Archibald and Gillingham do not specifically discuss the adaptive expectations hypothesis, this pattern of a negative effect of current price and a positive effect of price change is exactly what one would expect to find during a period of rapid price changes under an adaptive expectations hypothesis. It is as if people do not expect that recent rapid price increases will be sustained into the future. In terms of current-period actual prices (the variable Archibald and Gillingham used), the current expected price can be expressed as

$$P_t^* = P_t - r(P_t - P_{t-1}) \text{ when } P_{t-1} = P_{t-1}^* .$$

Thus, the effects of price level and price change should have opposite signs and the absolute value of the latter should be smaller that that of the former. Archibald and Gillingham found that the coefficient for (the natural logarithm of) gasoline price level was -.549 while that for price change was +.142. These coefficients imply a level of r of roughly 25 percent, which means that consumers increase their expectations of prices by about 75 percent of the discrepancy between last quarter's expected prices and current prices.

Before concluding that the evidence supports the idea that consumers react to both price levels and price changes, we should note that Archibald and Gillingham's findings are subject to alternative interpretations. Their sample period ended in the first quarter of 1974, exactly the period in which the effects of the 1973 oil

---

implies serial correlation of the errors and the partial adjustment mechanism does not. In practice it is difficult to tell if any observed serial correlation is due to the incorporation of the Koyck transformation or is inherent to the original data.

embargo were being felt. This timing, combined with the fact that the investigators pooled four quarterly observations per household to form their sample, means that the change measure they used for gasoline prices coincided with an oil embargo during the measurement period. If the price-change variable is picking up the effects of the embargo, the implication might be not that consumers respond to price changes as well as levels, but that they are unwilling to reduce consumption when the price level is due to a (perhaps temporary) supply disruption.[7]

The weight of the evidence is that consumer response is partly determined by the rate of change of prices, not just by price levels. The issue can be very important in periods of volatile prices. Much more work needs to be done to estimate the magnitude of consumer response to the rate of change of prices, and better data are needed for that work.

IS THERE A THRESHOLD FOR THE RESPONSE TO PRICE?

Voluminous psychological literature--mostly from laboratory studies but also from field studies of organizational behavior--supports the hypothesis that energy users often act as problem avoiders (Stern and Aronson, 1984) and the related hypothesis that responses to price may not occur until a threshold has been crossed. The notion of problem avoidance also suggests that once a threshold is crossed and the consumer believes that a response is needed, the size of the short-run adjustment may not be strongly affected by the size of the price change. The

---

[7] There are further complications with the Archibald and Gillingham study that cloud interpretation. Cross-sectional demand analyses are usually interpreted as long-run in nature, but Archibald and Gillingham interpret them as short run. They justify their interpretation on the basis of controlling for a vehicle stock measure and some locational measures through which the long run is supposed to operate. They cannot, however, control for all factors associated with the long-run response, and notably missing from their control variables is distance of residence from work. Other analyses have shown that work distance is a major determinant of the long-run gasoline demand response; thus it is not clear how to interpret Archibald and Gillingham's model.

findings of an experiment with time-of-use electricity
pricing are consistent with this view (Heberlein and
Warriner, 1982): households shifted electricity use to
off-peak periods in response to price differentials rang-
ing from 2:1 to 8:1. Within that range, however, attitudinal variables accounted for much more of the variance
in energy use than prices did.

It is possible to conduct studies to formally investigate the threshold hypothesis with respect to energy
demand. Experiments with alternative electricity rates,
such as the time-of-use pricing experiments, could be used
as tests if properly designed rate schedules were included
among the experimental conditions. It might also be possible to illuminate the issue through analyses of panel
data on electricity consumption because electricity rates
often change in different proportions in different utility
service areas.

DO PRICE INCREASES AND DECREASES
PRODUCE DIFFERENT RESPONSES?

There is evidence from psychological research that people
do more to avoid a loss than to reap an equal gain
(Kahneman and Tversky, 1979); this effect is sometimes
true for energy-related behavior as well (Yates, 1982).
This loss avoidance may mean that energy users will
respond more vigorously to a price increase than to a
comparable decrease. This possibility might be addressed
by analyses of panel data in the face of price increases
and decreases (as there have recently been for gasoline
and home heating oil).

CONCLUSIONS

Most energy policy models, including several used by the
Department of Energy's Office of Policy, Planning, and
Analysis, use one or two energy price measures that are
allowed to affect predicted energy demand via some simple
lag structure. In contrast, our examination of the price
variable suggests that at least three dimensions of price
may have important effects on energy demand. While in no
case is the empirical or theoretical evidence sufficient
to suggest concrete changes to existing policy models,
some general conclusions do seem justified.

Models dealing with energy sources that are sold under complex rate schedules should make allowance for differential response to average and marginal prices since both seem to affect demand. Similarly, if policy models are to be used to assess the short-term impact of pricing policy (as opposed to the long-run equilibrium effects), they must be reformulated so as to reflect the complex nature of response to both price levels and to rates of change in prices. Simple geometric lag structures cannot be expected to capture the dynamic aspect of real-life consumer behavior. Similarly, the dynamics of consumer response are also likely to be affected differentially if real energy price changes are the result of changes in nominal energy prices or of changes in the nominal prices of other goods.

We should note that all the hypotheses discussed in this chapter can easily be represented in terms usable in formal demand models. In fact, given sufficiently detailed data, these hypotheses can also be tested by modeling techniques. Thus, both problem-oriented research and general data collection efforts can feed into the model-building process in addressing the questions raised in this chapter. (Chapter 3 raises some questions about the nature of consumer response to financial stimuli that are not so easily incorporated into existing models.)

Although the several price variables discussed in this chapter can readily be modeled, there is not yet enough empirical evidence to offer much guidance on what changes to make in the models. This state of affairs can only be remedied by further analysis of existing data, by collecting new data, and by developing the theory.

On the question of average versus marginal prices, further analysis of existing data may yield enough information to reformulate current policy models in a more realistic manner. Any data base sufficiently rich to allow analysts to adjust incomes to account for the "income component" of rate schedules is also sufficiently rich to use equations that do not force the effects of this component to equal that of ordinary income. Very few demand analysts have reported the results of such model specifications; more should do so.

Existing data are not rich enough, however, to change models to account for the different dynamics that may result from changes in real versus nominal price or changes in price levels versus rates of change in price levels. There has been insufficient independent variation in these dimensions of price to allow precise estimation of their respective effects. In most data for any particular economy, real prices have been highly correlated with nominal prices. What variation does exist is confounded by the fact that effects that result in inflation in a particular period may have independent effects on energy demands. Furthermore, most analysts have access only to national cost-of-living indices and thus cannot capitalize on what cross-sectional variation there is in price levels. Similarly, there is very little cross-sectional variation in price changes for several important fuels (e.g., gasoline and fuel oil), and past price levels are negatively correlated by definition with rates of change in prices.

Because different features of price have covaried in the past, simple assumptions, such as that consumers respond only to real prices, may be reasonably adequate for modeling existing data--even if the assumptions are in error. But to the extent that future events do not follow past patterns, more differentiated knowledge about how price affects demand will have practical importance.

Some relatively inexpensive improvements to existing data sets might help analysts in addressing the price issues. The Bureau of Labor Statistics computes, but until recently did not release, a consumer price index (CPI) for each of 23 Standard Metropolitan Statistical Areas. If detailed geographic area measures were made available, the full set of these CPI measures could be used to compute price indices that are comparable both across area and over time. There may be enough independent variation in these data to determine the relative effects of real and nominal prices on energy demand. Furthermore, with data on the individual commodity prices upon which the indices are constructed, repeated measures of consumption over the same units of time--such as is available in a crude form from the Panel Study of Income Dynamics and may in the future be available from the Residential Energy Consumption Survey (RECS)--might lead to real progress in identifying the dynamics of consumer response to price levels and changes.

New and continued data collection will be necessary to provide satisfactory evidence about which dimensions of price affect consumer behavior. The past decade has presented analysts with a number of natural experiments involving energy price changes. Had the decade begun with a measuring instrument such as RECS in place, far more detailed information would be available for modeling energy demand behavior. If detailed measurement had been done before the change in economic conditions, each household could have acted as its own control for the analysis of the effects of the change in conditions. The lesson for the future is that it is necessary to have a data collection capability in place well before it is of any current interest. Furthermore, once a sample is created, it is valuable to maintain it by contacting sampled households on a regular basis. In short, to develop an adequate understanding of how price changes affect consumer behavior, it is important to conduct RECS and similar studies even during rather "dull" periods in which energy supplies are plentiful and prices stable.

3

The Effects of Financial Incentives on Energy-Efficient Investments in Residential Buildings

A number of studies have concluded that rates of investment in energy efficiency are far from economically optimal levels.[1] Substantial investments are not made that would, by substituting technology for energy, lower overall costs for energy users without changing their levels of services (Hirst, Fulkerson, Carlsmith, and Wilbanks, 1982; Office of Technology Assessment, 1980, 1982; Ross and Williams, 1981; Sant and Carhart with Bakke and Mulherkar, 1981; Solar Energy Research Institute, 1981; Stobaugh and Yergin, 1979). A study by a group at the Mellon Institute (Sant, 1979) concluded, for example, that the actual mix of fuels and capital characterizing energy consumption was quite different from what it would have been if society had been minimizing long-run costs: in 1978, the nation actually consumed 79 quadrillion Btus (quads) of energy, of which 45 percent was oil; in contrast, the Mellon study estimated that if consumers had been minimizing life-cycle costs, total consumption would have been about 59 quads and oil consumption would have been as much as 30 percent less. In addition, the total cost to society of energy services (fuel plus capital) would have been 17 percent lower in the model's least-cost case than in the actual 1978 situation.

---

[1] "Economically optimal" has been variously defined, but the essence of the concept is minimization of cost. The concept usually implies comparing the net present values of alternatives under stated assumptions about the operating costs of equipment, the discount rate for future energy costs, and the useful life of the equipment. Calculations of optimality should also discount future values according to the uncertainty that they will be achieved,

In the residential sector of the economy, rates of investment in energy efficiency are probably even further from optimal levels than in other sectors. Most home-owners are less likely than industrial or commercial decision makers to spend money on energy efficiency in exchange for cost savings over the long term (Hausman, 1979). It has been estimated that economically justified energy savings of 50 percent and more are technically possible in the average residence (Solar Energy Research Institute, 1981). Yet national tax return and Energy Information Administration (EIA) data show that only about 5 percent of all households made conservation investments in 1977 and 1978. Most households in the EIA survey invested about $720--about one-half of the level of $1,500 suggested by some analyses as optimal (California Public Utilities Commission, 1980; Energy Information Administration, 1980; Hirst, Goeltz, and Manning, 1982).

A number of explanations have been offered for why households invest less in energy efficiency than is economically justified. Some of the major barriers to these investments include:

- lack of information concerning conservation technologies and the magnitude of the costs and benefits of using them;
- the existence of confusing and conflicting information about the effects of energy-saving technologies and practices;
- limited choices because intermediaries (such as appliance manufacturers, building owners, or home builders), who are not concerned with operating costs, make many decisions;
- the relatively time-consuming and complicated processes necessary for one to become informed about appropriate conservation measures and to actually make purchases;
- consumer distrust of the suppliers of conservation services and information, particularly small contractors in the home improvement field;

---

but most economic analyses of energy efficiency have discounted these values at the market rate of interest, without considering that returns on investments in energy efficiency may be less certain that returns on investment in organized markets (Chernoff, 1983).

• the "invisibility" of energy flows; that is, the difficulty of observing and assessing the effects of one's investments or behavior on one's energy use; and
• lack of readily available cash to pay the high first costs of installation, coupled with an inability to find financing or a preference not to use it.

Because of the variety of recognized barriers to economically justified investment in household energy efficiency, programs aimed at reducing those barriers might be expected to succeed in shortening the time scale for replacement of inefficient capital stocks and in raising the final level of penetration for energy-efficient technologies. Governments and utilities are spending large sums on incentive-based programs to achieve these goals. The Tennessee Valley Authority, for example, has provided more than $250 million in interest-free loans to its customers (Berry, 1982), and federal energy tax credits are expected to cost $2.5 billion between 1981 and 1986 (Hirst, Goeltz, and Manning, 1982). A variety of smaller grant, loan, and rebate programs exist to encourage more investment by offering financial incentives.

Formal energy demand models should, ideally, provide a means of comparing the effects of alternative policy actions on investments in residential energy efficiency. Such models begin with a baseline estimate of the time scale and penetration level for new energy-efficient technologies without incentive policies (see Figure 1). Then, expected diffusion curves for different types and levels of incentives can be estimated and compared with the baseline. Analyses of this type have been made in formal models.

As we noted in Chapter 1, the methods used to define the baseline diffusion curves and the curves resulting from incentive policies inspire little confidence. There is usually no basis for validating the correspondence between a model and actual behavior. Little is known about either the dynamics of the adoption of energy-efficient technologies or about the final level of their penetration.

There is also little behavioral knowledge about the way incentives operate on which to estimate how incentive policies would change consumer response. In estimating incentive effects, analysts typically interpret each contemplated incentive as a change in the cost of energy or the cost of equipment. Such conversions involve assumptions that are often left implicit. For example, the

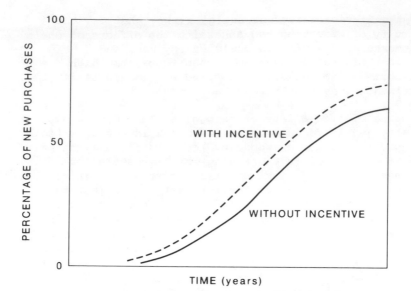

FIGURE 1 Hypothetical curves showing penetration of new
energy-efficient technologies as a function of time, with
and without a financial incentive.

Hirst-Carney (1978) model of residential energy use, in
analyzing the impact of federal tax credits, assumed that
the sole effect of the tax credit was to reduce the
initial cost of the equipment: that is, the 15 percent
tax credit was treated as a 15 percent reduction in the
cost of eligible energy-efficient devices. No consideration was given to other plausible effects of the tax
credit. For example, the fact of government support may
have bolstered some consumers' confidence that the covered
conservation investments would be effective. Other consumers may have largely disregarded the credit because
repayment would occur 4 to 16 months after purchase,
because they have no tax liability against which to claim
a credit, or because they do not habitually keep the
records or do the extra paperwork necessary to claim a tax
credit. To raise such possibilities is to suggest that
the economic meaning of a financial incentive is not
always obvious and to note the empirical questions that
must be answered if formal models are to offer reliable
interpretations of the effect of financial incentives.
     In exploring the relationship of financial incentives
to investment in the energy-efficiency of residential

capital stock, this chapter focuses on three issues: (1)
the relationship between the size of an incentive and consumers' responses; (2) the possibility that people may
respond differently to different types of incentive, even
when the incentives have the same financial value; and (3)
the importance of nonfinancial variables to the effectiveness of financial incentive programs. We discuss evidence
that the effect of incentive size may not be what is usually modeled and that incentive type and other nonfinancial variables may be critical to the success of incentive
programs. Finally, we present approaches for improving
understanding of how financial incentives work--or fail
to work--in the context of residential energy conservation.

HOW DOES THE SIZE OF AN INCENTIVE
AFFECT CONSUMERS' RESPONSES?

The simplest interpretation of a financial incentive is
that it lowers or defers the cost of energy efficiency
relative to the alternatives. Larger incentives make
investment economic for more people, so more people
ultimately make investments. Incentives may also increase
the pace of investment by making capital more available.
In this interpretation, the result is that both the pace
and the final level of investment will appear as smooth
functions of the size of the incentive. But incentives
may operate in other ways. They may, for example, function mainly as attention-getting devices; or they may
mainly speed action by people who would have taken the
same action without the incentive.

Incentives as Attention-Getting Devices

There is a rationale in psychology for thinking of an
incentive as a way to attract attention. If energy users
are problem avoiders who do not normally pay attention to
energy (Stern and Aronson, 1984), an incentive will matter
only if its size or the accompanying marketing effort is
great enough to command a consumer's attention. After the
incentive passes this threshold, further increases will
make little difference. In this interpretation, behavior
is a step function of the size of the incentive. The
aggregation of the individual step functions may produce
a smooth curve, but if the underlying process is atten-

tional, thinking of the size of the incentive as crucial misses the point: efforts to attract attention may make more difference than the size of the incentive.

There is some evidence that threshold effects do exist with energy use. Analysis of data from the Wisconsin time-of-use electricity pricing experiment (Heberlein and Warriner, 1982) showed that, over a range of price differentials from 2:1 to 8:1, the effect of price was not as great as that of personal commitment or other nonprice variables. Most of the effect of time-of-use prices on behavior seems to have been achieved by the 2:1 price differential, and additional price doublings made little difference. These findings, however, may not generalize to incentives aimed at major investments rather than at changes in the way consumers operate household equipment.

The only data on threshold effects for such incentives come from studies of consumer response to real or proposed loan subsidies. Those data suggest that there is little consumer interest in loan subsidies, except for interest-free loans. At least 70 percent of all home energy improvements are undertaken without the aid of loans (Berry, 1982), and offers of reduced-interest loans have not attracted many additional householders. In one study of participants in a conservation program that offered bank loans at reduced rates, 70 percent paid cash for retrofits averaging over $1,100 in cost (Stern, Black, and Elworth, 1981). A survey by Northeast Utilities (1981) reported that the amount of money people intended to borrow did not change as fast as the loan rate. But the availability of interest-free loans alters the picture. Programs that offer interest-free loans generally have a higher proportion of participants using loans than programs that offer only low-interest loans (Berry, 1982). In the study by Stern, Black, and Elworth (1981), 49 percent of people who did not have retrofit work done under a conservation program that offered reduced-interest loans said they would have taken advantage of interest-free loans (payable on resale of the house).

These findings are suggestive, but not conclusive. If large-scale loan subsidies are being considered, it would be worthwhile to be more certain about the possibility that a threshold effect exists. Research should specifically address the possibility of a discontinuity in consumers' responses to loan rates, such as might result from a process of attracting attention. This possibility can be studied initially in small-scale experimental studies that assess consumers' interest in loans at various rates.

Some such studies have been conducted (see Berry, 1982 for a review), but they have tended to be larger than necessary in their sampling effort and inadequate in the detail in which they explore variations in the incentives. Better methods for evaluating the effect of loan rates exist and are discussed in a somewhat broader content in the next section.

Incentives as an Impetus
to Speed Changes Already Planned

It is also possible that incentives affect only those energy users who are already paying attention to the costs of energy efficiency--that is, those who have already decided to invest. If so, incentives may increase the pace of investment, but not its final level. They might do even less: an incentive program may simply shift the financing of planned investments from savings or other sources of loans to the incentive program.

The limited evidence that is available is consistent with the hypothesis that incentives affect mainly the committed investors: 60 percent or more of the people who claim conservation and solar tax credits report that they would have invested as much without the incentives (Berry, 1982). People who take advantage of loan subsidies have similar characteristics to people who invest without special incentives: people with higher incomes are more likely to participate in subsidized loan programs and apply for larger loans (Berry, 1981; Northeast Utilities, 1981); people who live in energy-inefficient houses, and therefore have more need to invest, are more likely to take advantage of incentives for investment (Pacific Gas and Electric Company, 1982). Thus, it is likely that most people who take advantage of financial incentives would have invested anyway, sooner or later.

Incentives do appear, however, to increase the pace of investment: 78 percent of participants in a Pacific Gas and Electric Company interest-free loan program stated that they would not have installed their conservation equipment during the time period in which they actually purchased it if the loan had not been available (Pacific Gas and Electric Company, 1982). Similarly, only 29 percent of householders who accepted low-interest conservation loans from Minnesota's Northern States Power company said they would have made the same investments without the loan (Hirst, Goeltz, Thornsjo, and Sundin, 1983). The

clearest evidence comes from a 2-year study of participants in the Bonneville Power Administration's home weatherization program (Hirst, White, and Goeltz, 1983b):
households that took advantage of the program's interest-
free loans reduced their energy use by 12 percent in the
first year and by 14 percent in the second year as compared with nonparticipants; households that used the program's energy audits but did not take the loans showed no
savings in the first year but 8 percent savings in the
second year, as compared with those who did not use the
audits. The incentive seems clearly to have speeded
investment; only further follow-up research can determine
if there was also an effect on the final level of investment achieved.

The idea that incentives work only for people who are
already paying attention to energy efficiency is plausible
because of the kind of behavior involved. Consider the
difference between a low-interest car loan and a low-
interest energy loan. The purpose of the former is
usually to influence a choice among alternatives: most
people who are considering buying a car are planning to
borrow, so are likely to shop around for a low loan rate.
A loan subsidy may get such people to purchase a different
model or to make a planned purchase sooner; it is unlikely
to increase the number of cars in use. But the primary
purpose of a conservation loan program is to spread an
innovation--to increase the number of people who decide
to make a purchase. Thus, energy loans probably are not
perceived and do not function like car loans. Someone who
has not yet decided on energy efficiency is not likely to
be shopping for low loan rates and will not notice them.
Furthermore, a conservation incentive program that says
one must go into debt to conserve energy is not making an
effective sales pitch, regardless of the interest rate
offered.

The idea that incentives only affect people who would
have invested anyway remains only a hypothesis. But
because of its policy implications if true, it seems
advisable to conduct some small experiments to test it.
Research designs that include comparison groups not
offered incentives are the best way to control for self-
selection--the possibility that people choose to participate in an incentive program only after making the decision to invest. A comparison group makes it possible to
separate differences due to an incentive from those due
to other, simultaneous events. It also makes it possible
to tell whether people offered the incentive differed in

any potentially important ways from those not offered the
incentive. Still more reliable interpretations can be
made from randomized experiments--those in which households are randomly assigned to treatment groups. This
approach puts statistical bounds on the possibility that
treatment groups differ systematically in ways (including
unmeasured ones) that might affect their response to an
incentive. Comparison groups not offered incentives can
be obtained by studying programs in nearby areas (e.g.,
different utility service areas) that do and do not offer
incentives. But because the programs (and areas) may
differ in other ways than the incentives they offer (see
below), random assignment of eligible consumers to different types of incentive is the surest way to obtain reliable data.

Small-scale experiments with incentive programs can be
designed to test simultaneously the effects of incentives
and other program components. For example, such experiments could explore the possibility that for encouraging
new investments, a conservation program might get more out
of its resources by simplifying the process of investment
than by offering a financial incentive.

HOW DOES THE TYPE OF INCENTIVE
AFFECT CONSUMERS' RESPONSES?

Formal demand models usually compare types of incentives
in terms of their economic value, and, given the present
state of knowledge, the comparison requires making tenta-
tive assumptions. For example, a simple economic model
may compare a loan and a rebate in terms of net present
value: rebates decrease present costs while loans defer
them, so, given knowledge of consumers' discount rates,
the two incentives can be compared by calculating the
present value of the loan. By this method it becomes
possible to tell which loan rates are equal in economic
value to which rebate rates. Thus, to equate loans or
rebates in terms of economic value, the relevant discount
rate must be postulated or estimated.

The available data on consumer preference for incentives, if interpreted in the language of discount rates,
suggest an anomaly. Preference for loans or rebates
appears to be a function of income, with higher-income
consumers more likely than lower-income consumers to pre-
fer loans (Berry, 1982). If this difference in preference
is interpreted in terms of net present value, the implica-

tion is that high-income people, who express a preference to defer some present cost with a loan, have higher discount rates than low-income people. This finding appears to contradict other evidence that income is inversely proportional to discount rates (Hausman, 1979; McFadden and Dubin, 1982). The contradiction can probably be resolved by focusing on attributes of the types of incentive other than those related to time discounting. Perhaps rebates are attractive to low-income people because they reduce the perceived size of the capital commitment while loans do not. Or it may be that low-income consumers are simply averse to indebtedness.

Incentives vary qualitatively in several ways that are not captured by the usual calculations of economic value. The kinds of incentives offered for energy-efficiency investments include grants, price discounts, rebates, and loans. A qualitative classification of types of incentives may prove useful for understanding energy users' responses.

Grants decrease the cost of an investment before it is paid and are usually offered by a government agency. Some of them, such as those offered by low-income home weatherization programs, cover labor as well as capital costs. Including labor may attract energy users because it solves the problem some people have of finding someone to do the work, but it may repel people if they do not trust the program or the workers it provides.

Discounts on the price of energy-efficient equipment are usually offered by manufacturers or retailers. They are like grants in terms of their immediacy and may be even easier to obtain because they do not involve paperwork.

Rebates are delayed discounts. They usually involve a process of filing a claim, which can vary from trivial to tedious. Their effectiveness is likely to depend on whether claiming the rebate is seen as worth the effort.

Tax credits, like rebates, are delayed discounts, but they are different from rebates in that they have longer delays and require record-keeping as well as filing of forms. They require a change in the routine of filing tax returns and are not available to energy users who have no tax liability.

Loans must be approved in advance, and the possibility of refusal may deter some consumers from applying. In addition, people may think about loans differently from the way they think about other incentives: for a person with little economics education, discounts, rebates, and even tax credits are probably easier than loans to imagine as price decreases. If so, a smooth curve relating the size of an incentive to behavior will be observed more regularly for nonloan incentives than for loans. Furthermore, there is evidence that many householders simply do not consider going into debt for energy efficiency. Householders consistently say that the availability of low-interest loans makes little difference in their decisions to invest in home energy efficiency (e.g., Olsen and Cluett, 1979; Stern, Black, and Elworth, 1981). Interest-free loans, as already noted, may be qualitatively different from low-interest loans if they do not seem like borrowing. Generally, procedures that make loan repayment seem painless--by deferring it until a house is sold or by ensuring that loan payments are never greater than energy savings--may make a large difference in response to loan programs.

The above descriptions emphasize certain qualitative characteristics of financial incentives that seem likely to make a difference in consumer response: delay in receipt of the incentive, effort involved in obtaining it, uncertainty that it will be received, the incentive's effect on household budgets and cash flows, and trust in the offering organization. We identify these particular variables as important on the basis of examination of behavioral research on energy conservation programs, on energy use more generally, and on other actions of individuals and households (for a detailed discussion, see Stern and Aronson, 1984:Chapters 3, 4).

Some of the above variables seem capable of quantification and inclusion in formal demand models. Delay, for example, differentiates tax credits from more immediate incentives and may differentiate among programs offering the same type of incentive (e.g., rebates). The concept of discounting can be used to estimate the effects of delay, although the lack of an independent empirical basis for estimating discount rates makes this a questionable approach until more data are in. Uncertainty about receiving the incentive differentiates among tax credits, loans, and rebates--and especially within the class of loan programs. This uncertainty can be estimated either from program data or from the perceptions of potential clients; the latter may be the more relevant index. Effect on budgets and cash flows distinguishes primarily among loan programs. For example, programs that defer repayment until a house is sold may be more attractive than those that defer payment for 7 years, even in an area where the average house is sold in 7 years.

We do not believe that sufficient data yet exist for quantifying even the most quantitative of these variables. But methods exist for gathering such data, and it is already possible to offer some hypotheses for testing: (1) Delay in receipt of an incentive is an especially important barrier for low-income consumers; (2) Programs that keep energy investments from draining household budgets are significantly more attractive than other programs offering incentives of similar size, but special marketing efforts may be necessary to emphasize this feature; (3) The red tape, record-keeping, and other effort needed to take advantage of some incentives is a significant deterrent to consumers, especially low-income consumers.

These and related hypotheses can be investigated. While some relevant data might come from evaluations of existing incentive programs and pilot studies of new programs under consideration, a more systematic and promising method can be used: small, laboratory studies that compare people's initial interest in a great variety of possible incentive packages. Methods of focused group discussion can provide some initial insights, but more quantitative estimates can be made by using methodology adapted from psychometric studies of decision making under uncertainty (e.g., Kahneman and Tversky, 1979; Tversky and Kahneman, 1981) or from research on multiattribute decision analysis (e.g., Klein, 1983; Svenson, 1979; Tversky and Sattath, 1979). In these approaches, people are presented with a series of hypothetical choices and decision problems that vary in terms of the attributes, trade-offs, steps, contextual factors, and rules used in making decisions. With these methods, a large number of theoretical questions about choice can be investigated in a few experiments.

Of course, such laboratory studies may not be generalizable to actual choices. Self-reported interest and intention are highly imperfect indicators of behavior in a new situation. But the purpose of laboratory studies is not to draw conclusions but rather to narrow the range of likely hypotheses for more rigorous test under field conditions. Small field experiments that systematically vary the factors that seem important would be a logical next step. In addition to or after initial field trials, the most promising incentive packages can be implemented experimentally to compare their effects with those of existing incentive packages or other comparison conditions.

Some qualities of financial incentive programs are less easily quantified, even with psychometric methods. Some of them, however, are equally likely to be important. One of these, trust in the sponsoring organization, has been mentioned above. It and others are discussed in the next section.

HOW DO CONSUMERS RESPOND TO NONFINANCIAL FEATURES OF INCENTIVE PROGRAMS?

Nonfinancial factors clearly have important effects in incentive programs. At times, they may make more difference than the incentives themselves. Conservation programs that are identical in terms of the incentives they offer are received very differently by their prospective clients. For example, New York State requires its regulated utilities to offer financing of energy efficiency at 9-11 percent interest; the proportion of households receiving energy audits who take advantage of the financing ranges from less than 1 to more than 40 percent across the utilities (Scherer, 1981). The Bonneville Power Administration offered interest-free loans through 11 utilities participating in its pilot weatherization program. The proportion of households given energy audits under the program ranged from 10 to 58 percent, and the percentage of audited households accepting the interest-free loans ranged from 8 to nearly 90 percent (Lerman, Bronfman, and Tonn, 1983).

Most of the available evidence for explaining this variation comes from studies of conservation loan programs. It suggests that the most important influence on a program's effectiveness is not the loan rate or the term but implementation: promotion to the clientele, success in simplifying the decision process and in alleviating fears of shoddy work, and the ability of the sponsoring organization to gain trust. The same variables seem to affect conservation programs that rely primarily on information, such as the Residential Conservation Service and other energy audit programs.

Promotion begins with getting a consumer's attention. Utility bill enclosures generally fail in this respect, while direct mail and media advertising are somewhat more effective; news items including statements from prominent public officials attract still more attention (Rosenberg, 1980). Personal communication from friends or trusted local organizations is probably the most effective form of promotion (e.g., Leonard-Barton, 1980, 1981; Rogers with Shoemaker, 1971), and this approach has been used with success in energy conservation programs (Olsen and Cluett, 1979; Fitchburg Office of the Planning Coordinator, 1980).

Promotion also involves an effort to "sell" conservation to households. A recent study of the pilot weatherization program of the Bonneville Power Administration (Lerman, Bronfman, and Tonn, 1983) concluded that differential success among the participating utilities was due in considerable degree to the level of a utility's effort and commitment to the program. In particular, utilities succeeded in getting more homes weatherized when they had larger staffs for conservation, when management and personnel were dedicated to the program, and when their staff members promoted the program rather than contracting the job to outside organizations that were paid a fee for each audit they conducted.

Simplification of the decision process is often identified as a major benefit energy conservation programs can offer households. Even though incentives directly address only financial issues, the success of an incentive program may hinge on how well it assists clients in resolving other problems. A home retrofit investment depends on choices about how to get information, what information to believe, what to do, how to get it done, and how to pay for it--and each of those choices can be difficult. The convenience some programs offer in terms of "one-stop shopping" for energy conservation has sometimes proved more important to potential clients than the availability of a financial incentive (Stern, Black, and Elworth, 1981).

Residential energy consumers are concerned about the reliability of the work done by providers of energy-efficiency home improvements. In one study of a statewide residential conservation program that offered low-interest loans, free audits, a pool of certified contractors, and inspection of all retrofit work (Stern, Black, and Elworth, 1981), the two reasons most frequently cited by participants for their involvement were: "I trusted the work because it would be inspected" (98 percent) and "I didn't have to worry about finding a reliable contractor" (96 percent). In contrast, only 14 percent of participants cited the availability of low-interest loans as a reason for their participation. Although consumer protection guarantees are not a part of financial incentives, they seem to make a major difference in the attractiveness

of a program that offers incentives. If participants'
self-reports are a valid guide, improving consumer
protection may be a more effective way to spend scarce
program funds than offering stronger financial incentives.

Finally, trust in a conservation program is obviously
crucial to its success. Trust probably makes more difference for programs that offer energy information or
installation services than for those that offer only
financial incentives, but since these features are so
often combined, trust can be a major issue in many incentive programs. Even in programs that offer nothing but
an incentive, trust in the sponsoring organization may
affect its ability to get consumers' attention. Some
suggestions for building trust in programs that involve
energy information are offered in Stern and Aronson
(1984).

Most of the above variables are not how readily quantifiable for inclusion in formal energy models--and they
may not be even after additional research. Existing
models generally do not use concepts that correspond to
those qualitative variables, which therefore usually show
up as variability in estimates of the rate and ultimate
penetration of energy efficiency. Given the extreme range
of variation among similar programs in these respects, the
qualitative factors are obviously candidates for further
analysis.

Such analysis is best carried out by problem-oriented
research. The importance of such factors as consumer
protection guarantees, simplicity of decisions, and the
identity of the organization offering the incentive might
initially be addressed by detailed small surveys, psychometric experiments, or even focused group discussions of
the type used in market research. But the limitations of
these methods are more severe here than for analyzing loan
packages, because it may be impossible to describe the
alternatives well enough for participants to understand
them. An alternative first step would be to address the
issues of promotion, implementation, trust, and so forth
in program evaluation studies. Particularly valuable are
studies of groups of programs (e.g., Stern, Black, and
Elworth, 1981): better yet, groups of programs that are
identical in their formal requirements but are implemented
by various organizations (for example, the Bonneville
pilot program studied by Lerman, Bronfman, and Tonn (1983)
and residential conservation programs operating under a
common set of state guidelines). Such research can set
approximate bounds on the size of effect a given incentive

may have depending on other conditions of its implementation. It can also generate hypotheses to test in small-scale field experiments aimed at finding the most effective ways to use the available resources. Similar research on programs that do not feature financial incentives (e.g., energy audit programs) can help identify non-financial interventions that might greatly improve the return on an expense for financial incentives (see Chapter 4).

CONCLUSIONS

Consumer response to different types and levels of financial incentives raises important policy questions both for governments and for the many gas and electric utilities that offer such incentives. The overall cost of federal and state residential energy tax credits, loan subsidies for residential retrofit, and rebates for retrofit and purchase of energy-efficient appliances is probably several hundred million dollars a year. Unfortunately, very little is known about the absolute and relative effectiveness of these financial incentives in stimulating installation of energy-efficient technology. Underlying this ignorance is a lack of knowledge about how incentives affect consumer behavior.

Financial incentives for residential energy efficiency seem to have positive net effects, at least under some conditions. They can accelerate the rate of investment, but their effect on the final level of penetration or investment has not yet been established. The magnitude of this effect is difficult to establish because there are no reliable estimates of the final level of investment that would be observed in the absence of incentives. Also, the level of investment depends simultaneously on incentives and on other factors that are influenced by policy.

Data can be collected to improve the basis for estimating the penetration of energy-efficient technology in the absence of incentives. Such data could come from large-scale surveys of existing capital stock and of rates of replacement of the equipment. Large panel surveys could determine rates of replacement, but to assess the nature of the new stock with respect to energy efficiency, smaller, more detailed surveys might be a more efficient research method. The existing literature on patterns of diffusion of innovation can generate hypotheses to test against survey data, and the data could be used to estimate the baseline for penetration of new technology.

It seems clear that the effect of incentives is very much dependent on factors not usually incorporated in formal demand models. Some of these, such as threshold effects, the income level of potential participants, length of delay in realizing the incentive, and the effect of an incentive on household cash flow, are fairly easy to quantify and include in formal models. But promotion and management of incentive programs, two variables that seem among the most critical (Lerman, Bronfman, and Tonn, 1983; Scherer, 1981), are not so readily quantifiable. Furthermore, the effect of an incentive seems to depend on the availability in or around the program of other aids to consumers, such as credible energy information, assistance in simplifying the decision process, and consumer protection. Trust in the organizations responsible for an incentive program is an important qualitative factor, and its importance probably increases if the incentive program also offers information or installation services.

Because so little is known about the magnitude of the effect the above variables have on consumer behavior, one cannot now predict or explain consumer response to different kinds of incentives (interest-free versus low-interest loans, rebates versus loans, and so forth). But as this chapter notes, it is possible to conduct problem-oriented research in order to produce estimates, or at least upper and lower bounds, for the effects of many of the important quantitative and qualitative variables. The existing data, inadequate as they are, raise several questions that should be the focus of research:

* Are partial loan subsidies of any value at all in encouraging new investment in conservation or in accelerating planned investment?

* Are rebates, grants, or other immediate incentives especially valuable attention-getting devices? Are they the only forms of incentive likely to affect low-income people?

* Is some of the money spent on financial incentives better spent on nonfinancial modes of attracting investment, such as marketing of conservation programs, guarantees of consumer protection, simplification of investment and payment procedures, and liaison between organizations providing incentives and others that may have higher credibility as information sources?

Our analysis suggests that a small proportion of the resources devoted to offering energy tax credits, utility loan subsidies, and other financial incentives could usefully be redirected to a relatively inexpensive effort to conduct studies addressed to the above questions. Such studies hold the potential not only for evaluating existing incentive programs, but for advising decision makers on how best to spend moneys allocated to promoting investment in residential energy efficiency.

The available formal models do not now encompass many of the important variables discussed in this chapter. These variables are not easily quantified, and they interact in unknown ways, making the analytic problems very complex. Some models are particularly inappropriate for analyzing the effect of financial incentives. Among these are the simple (e.g., reduced-form econometric) models of household energy demand that do not even include capital costs within the model, but estimate demand only from fuel prices and incomes. Other, more disaggregate, models explicitly include household choices regarding purchase of new equipment (Dubin, 1982; Goett and McFadden, 1982). These equipment choice models are explicitly sensitive to operating cost (which is a function of energy efficiency) and capital cost (which is affected by financial incentives). Such models are better able to incorporate some of the variables discussed here. A brief discussion of how one might do this appears as Appendix A. However, because of the importance of qualitative factors, it makes sense to be modest in the use of models for analyzing incentives for energy efficiency.

High priority should be given to investing analytic resources in studies of the organizational and behavioral factors that seem often to spell success or failure in incentive programs. When one interest-free loan program has only 8 percent of its eligible households using the loans while a formally identical program has 90 percent participation, a research program restricted to analyzing the readily quantifiable financial variables--in formal models or by other methods--is ignoring the factors that may prove to be the most important.

4

The Effects of Information
on Energy-Efficient Investment

Lack of adequate information is frequently blamed for the
slowness of consumers' responses to the rapid price
increases for energy over the past decade. The assumption of economic rationality generally followed in formal
energy demand analysis emphasizes the role of information:
under that assumption, rational action requires full
information about the available alternatives and their
relative costs and benefits. Energy prices provide some
of this information, since they show the economic effect
of using different amounts of different fuels. But prices
do not provide information about the effects of particular
actions on energy use. If consumers do not know how much
energy they can save with attic insulation, or that checking for leaks in industrial boilers can cut a business's
energy use by 10 percent, or that two refrigerators of the
same size may take vastly different amounts of electricity
to run, they will fail to take actions that would save
them money.

In a period of rapidly changing energy prices and of
threats to the availability of fuels, it is important for
energy users to know how their choices will affect their
energy consumption. To help consumers make choices in
their own interest, government has developed and dissemi-
nated huge amounts of information about the effects of
actions on energy use. It has calculated the fuel economy
of automobiles and distributed millions of copies of the
ratings to prospective automobile purchasers; it has
sponsored programs involving energy audits of homes and
nonresidential buildings; it has supported research on the
effects of new energy-saving technologies on energy use;
and it has published reports of the findings of such
research. This information can make a difference. The
automobile fuel economy labels required by the Environ-

61

mental Protection Agency, for example, were rated the most useful source of gasoline mileage information by new car purchasers in 1978 and 1979, a period in which more than two-thirds of purchasers rated fuel economy as a "very important" or "extremely important" factor in their purchase decisions (McNutt and Rucker, 1981).

Because information is necessary for an economically rational response to energy price signals, because government has relied heavily on information to promote energy efficiency, and because information can be an effective policy tool, energy demand analysis that cannot assess the effects of information is seriously incomplete. This chapter summarizes existing knowledge about what information energy users have and about how policy makers can provide better information for consumers; it also discusses two conceptual approaches for analyzing energy demand based on a recognition that energy information may be incomplete.

HOW COMPLETE IS CONSUMERS' INFORMATION?

In formal demand models, the simplest assumptions about energy information are that full information is available to consumers and that they use it: rational economic action follows. The converse is also true: models that assume rational economic action also assume, implicitly or explicitly, that energy users are acting with full information.

But full information is not always available. The economic cost of future energy is unknown, and even the effects of action on energy use are often uncertain. In commercial buildings, for example, actual energy savings from particular sets of technical improvements are often 50 percent more than predicted or 80 percent less than predicted from the best available technical information (Office of Technology Assessment, 1982).

Furthermore, the information that is available is not necessarily used when energy users act. Evidence shows that people are often unaware of available information; when they are aware of it, they often fail to act on it because of inertia, lack of comprehension, mistrust, or some other reason (for a discussion of the evidence on this point, see Stern and Aronson, 1984). To cite one of many examples, home energy audits have been offered to residential customers at low cost or even for free, but typically fewer than 5 percent of the eligible consumers request the audits. And of those who have had energy audits, only a minority take energy-saving action as a result. There are several reasons for this, mostly apart from the accuracy of the information: people may not trust the source of the information or the ability of contractors to do the work well; they may find it too difficult to gather all the information necessary to decide which investments to make; or they may not have the necessary capital to invest in energy-efficient technology.

Such problems with energy information constitute a major barrier to the penetration of energy-efficient technologies. Although it is impossible to estimate the size of the barrier with precision, two kinds of findings offer a rough gauge of its magnitude. Numerous experimental studies of one informational procedure, feedback on energy use, find that it yields average reductions in energy use of around 10 percent--and up to almost 20 percent when energy costs are high--without any investment in technology (Winkler and Winett, 1982; for a review of the literature, see Geller, Winett, and Everett, 1982: Chapter 5). And studies of residential energy conservation programs that act largely by offering information show that under some conditions participants invest twice as much in energy efficiency as nonparticipants during the same period (Stern, Black, and Elworth, 1982a) or save twice as much energy (Hirst, White, and Goeltz, 1983b).[1] Thus, it is safe to assert that improved delivery of information could make up a significant part of the difference between the level of adoption of energy-efficient technology under present conditions and the level of adoption that is economic for energy users. Information could certainly speed the rate of adoption, and it might also bring the final level of adoption closer to the economic optimum.

Treatment of Information in Formal Models

Formal demand models have little to say about what people do with incomplete information; how they respond to different kinds of information; or how they deal with an environment in which information is confusing, conflict-

---

[1] It is difficult to conclusively evaluate what proportion of this effect is attributable to the self-selection of program participants.

ing, and often untrustworthy. Instead, formal demand models ordinarily subsume the effects of information under other explanatory concepts. For example, lag coefficients and other indices of investment dynamics can be interpreted in part as mathematical expressions of the slow spread of full information among energy users; whether or not they are so interpreted, such a process shows up as a dynamic effect in models. If information speeds adoption, it changes such lag coefficients.

Another example is the concept of the discount rate (discussed in Chapter 1). When estimated from data on purchases of energy-efficient equipment, the discount rate reflects not only a preference for present value over future value, but energy consumers' responses to imperfections in the available information and failure to believe and act on information that may in fact be accurate. Improvements in information or in consumers' use of it would then appear to change the discount rate in models that estimate that variable.

Knowledge About Information
From Problem-Oriented Research

Surveys of energy users' beliefs and knowledge, small-scale experimental studies involving energy information, and formal evaluation studies of information-based energy programs offer some insights about what information consumers have and how informational policies and programs can change what energy users know and what they do.

The evidence is that energy users possess incomplete information for making energy choices. Furthermore, several studies have shown systematic misconceptions among energy users about the amounts of energy used by various household appliances (Becker, Seligman, and Darley, 1979; Kempton, Harris, Keith, and Weihl, 1982; Mettler-Meibom and Wichmann, 1982). People tend to overestimate the amount of energy consumed by household lighting and televisions and to underestimate the energy used by water heaters and some other appliances. Parallel misconceptions exist about what actions can save energy in the home: a survey of 400 Michigan families found that the average householder believed reduced lighting could save twice as much energy as reduced use of hot water (Kempton, Harris, Keith, and Weihl, 1982). If people act on such beliefs, the ultimate penetration of energy-efficient water heaters is likely to fall far short of what would be in homeowners' self-interest.

There is a pattern to misconceptions about home energy use. Generally, overestimation occurs for energy uses that are visible or that must be activated by hand each time they are used; underestimation occurs for energy uses that do not have these characteristics. This pattern follows what cognitive psychologists call the availability heuristic (Tversky and Kahneman, 1974): people tend to overestimate the frequency or importance of events that are easily called to mind.

It is worth emphasizing that people's misconceptions persist despite the availability of other, presumably more accurate, information. The misconceptions can persist even when accurate information is delivered directly to people who are misinformed. In the Michigan study, for example, householders who received computerized energy audits of their homes with specific recommendations for energy-saving activity had virtually the same patterns of belief about what saves energy as people who had not received the energy audits.

Improving Consumers' Information

At least one readily quantifiable factor is known to mediate the impact of information on action--the cost of energy (Stern and Aronson, 1984). This is best demonstrated by the research on the effect of frequent feedback about household energy use on future energy use (reviewed by Winkler and Winett, 1982). As noted above, energy savings achieved in feedback experiments are a function of the cost of energy to the household. When households have low energy costs, feedback has no effect, but with high energy costs, the savings have been as high as 20 percent. Corroboration of the relationship between price and information is available from research using an entirely different methodology. Using simple econometric models to estimate the effect on national energy consumption of expenditures by the U.S. Department of Energy on conservation programs, a group at Oak Ridge National Laboratory found that federal expenditures were significantly related to energy use only in interaction with energy price (Greene, Hirst, Soderstrom, and Trimble, 1982). One interpretation of these findings is that information matters only when prices are relatively high. The other side of this is more to the point at present, now that prices are high: the higher energy prices rise, the more important is the quality of the energy information that consumers have.

But to know that well-delivered information makes a difference does not tell how to deliver it well. The evidence from numerous program evaluation studies and controlled experiments shows that merely making energy information available to people is not enough: offering printed advice on how to save energy, for example, usually has little or no effect on behavior (for reviews of the experimental research, see Ester and Winett, 1982; Geller, Winett, and Everett, 1982; Shippee, 1980; Winett and Neale, 1979). The evidence suggests, however, that energy information can make a difference if it is presented in an attractive format and an easily understandable style, if it is vivid and personalized, if it is clearly relevant to the particular energy user's situation, if it is available through several media (prominently including word-of-mouth), and if it comes from a trusted source (for a discussion of the evidence on these points and the analysis leading to these conclusions, see Stern and Aronson, 1984).

Unfortunately for the purposes of formal policy analysis, none of these factors is readily quantifiable. Yet some of them can have a sizable effect on energy use. For example, people who viewed a videotaped information program constructed on many of the above principles saved 10 to 20 percent of household energy in comparison with control households (Winett et al., 1982). With additional information in the form of energy use feedback, energy savings increased to around 15 to 25 percent. These savings did not involve any investments in improved energy efficiency.

In energy audit programs, which use information to promote energy-efficient investments, the evidence suggests that a program's effectiveness depends on such qualitative variables as the program's promotional efforts, convenience and consumer protection features, and ability to gain trust (as discussed in Chapter 3). Other important qualitative variables include the effectiveness of energy auditors and other program personnel as communicators and various features of the ways recommended investments are described to the program's clients (for further discussion of these issues, see Stern and Aronson, 1984).

The net effect of programs for residential energy efficiency that rely largely on information can sometimes be substantial. One example illustrates the potential. A residential conservation program operating in the Northeast in 1979 offered free energy audits, access to

approved contractors, inspection of all installed improvements, and access to reduced-interest loans. Of homeowners requesting the energy audits, about 20 percent had
work done under the program, and they did virtually all
the work the auditors recommended as economically justified (Stern, Black, and Elworth, 1981). According to the
reports of those who requested audits but did not sign
contracts with the program, they made major investments
projected to save more than 50 percent of what the program
participants could be expected to save. By contrast, a
comparison group of eligible homeowners who had not
requested energy audits reported investments that would
save only about 25 percent of what the program's full
participants could be expected to save (Stern, Black, and
Elworth, 1982a). If these estimates are accurate, the
high energy prices of the late 1970s had motivated homeowners to make about one-quarter of the economically
justifiable investments in energy efficiency by mid-1980,
while people who requested audits from the conservation
program were making, on the average, more than half of
the economically justifiable investments.[2]

The results of this study suggest that a well-
constructed informational effort aimed at energy-efficient
investments can go a long way toward bringing about the
investment that full information would induce in rational
actors. The data as a whole suggest that the qualitative
aspects of information, more than the quantity of information available, is crucial to the effect of information

---

[2]This finding may be an overestimate of the effect of
information because the program offered financial incentives as well as information and because program participants were self-selected. But financial incentives probably account for only a small part of the program's effect
because less than one-third of the 20 percent who signed
contracts with the program used its low-interest loans
(Stern, Black, and Elworth, 1981). And self-selection may
also have been a minor influence. The sample of nonparticipants differed little from a statewide sample in an
adjacent state, either in socioeconomic characteristics
or in investments in energy efficiency. Furthermore, the
factors that influenced investments in the state that did
not offer a program accounted for only a small amount of
investment compared with what program participation seems
to have produced (Stern, Black, and Elworth, 1982a).

on action. But the data do not tell which qualitative
factors matter most nor how much difference each makes
alone or in combination with others.

HOW DOES INFORMATION AFFECT BEHAVIOR?

It may be reasonable to imagine the effect of information
on energy use in the same terms as as economic stimuli are
imagined to affect economic behavior: by producing an
eventual steady-state response after a period of dynamic
change. Thus, it makes sense to ask if information
affects the final penetration of energy-saving technologies and practices, the rate at which the changes take
place, or both. Although researchers have not framed the
questions in quite this way, the available knowledge suggests answers. To the extent that purchase decisions are
influenced by erroneous impressions about how much energy
a technology uses, information must affect the final
penetration of energy-efficient technology. And to the
extent that accurate information penetrates the society
slowly, the responses to changing economic conditions are
likely to be slowed. There are reasons to expect that
information can make a difference, then, in both the
statics and dynamics of energy demand.

To understand these effects, one must understand the
process by which energy information comes to affect behavior. There are at least two possible descriptions of the
process, each with its own implications for what knowledge
is needed: that people act rationally on the basis of
whatever information they possess; or that information
diffuses through informal networks with people acting by
example or on trust, rather than as a result of rational
choice.

Rational Action Based on Imperfect Information

As a first approximation of the role of information, formal demand analyses might suppose that consumers act
rationally on the basis of the information that is in
their awareness, as they understand, interpret, and trust
it. In this view, _available_ information is transformed
by social and behavioral processes into _effective_ information (that is, the information consumers act on);
behavior follows from effective information by principles

of rational choice.[3] Thus, the key analytical problem
is to learn what leads consumers to understand and interpret the available information as they do. Energy information may have static and dynamic components; consequently, the rate and final penetration of accurate energy
information set limits on the rate and final penetration
of economically rational responses to conditions in energy
markets. To understand the effects of informational
policies and programs, an analyst must determine how much
and how quickly policies and programs, operating in the
existing informational context, change effective information for energy users.

Given this "effective information" model, an obvious
first step is to gather information about what consumers
believe to be true about energy use and energy-efficient
investments. This step can be carried out by conducting
detailed surveys of consumers' beliefs about the costs and
benefits of various investments in energy efficiency
(e.g., Kempton, Harris, Keith, and Weihl, 1982). Data
from such surveys could be incorporated into existing
formal models by substituting estimates of perceived costs
and benefits for the terms that define the costs and benefits of alternative actions. One might compare an existing model that estimates energy-efficiency investments
under full information with a parallel model that holds
parameters constant but substitutes beliefs about capital
cost and operating cost for the best available information
about actual costs. This procedure could estimate an
approximate upper bound for the effect improved information might have on the rate and final penetration of
energy-efficient technologies. Data on consumers' beliefs
could also be combined with measures of actual energy use
and investments in energy efficiency to calculate discount
rates based on effective information. (This could be
done, for example, with discrete choice models of the type

---

[3]The assumptions of rationality may not hold, however.
By most definitions, a rational consumer is one who con-
verts beliefs about costs and benefits into expected value
(i.e., multiplies by probability), corrects for inflation,
compares the considered action with the costs and benefits
of alternative actions, and takes the effects of any tax
benefits into account in all these calculations. Some of
these assumptions are questionable, and others are
improbable for some consumers (see Chapter 2; Kempton and
Montgomery, 1982; Stern and Aronson, 1984).

presented in Appendix A.) A plausible hypothesis is that when investment choices are modeled on the basis of what people believe to be true about costs and benefits, rather than on the assumption of full and accurate information, estimated discount rates will come closer to those used by most investors in financial markets.

Research to assess consumers' beliefs should be disaggregate because beliefs about energy may vary geographically, by fuel, or as a function of sectors of the economy, the purposes for which energy is used, the consumer's economic status, or other factors. Such disaggregated research could identify for what groups misinformation is a serious barrier to energy efficiency and for whom special informational programs might be developed.

Surveys to assess effective information would be useful, but they do not address the key analytical question of how available information gets transformed into energy users' beliefs (effective information) or the key policy question of how to improve effective information. The answers to these questions are difficult to get and then difficult to incorporate into formal models. The following questions must be addressed: Which information do people notice? Which information do they understand? Which information do they believe? What makes information credible? How do people establish their beliefs in the face of conflicting information, and what information, if any, can change such beliefs? How do people account for the uncertainty of information? These questions have rarely been asked in the context of formal energy demand models; indeed, they call to mind processes of attention, cognition, and judgment that are not readily modeled as economically rational action. Yet just such processes may make the difference between a successful informational effort and a worthless one (Stern and Aronson, 1984).

What does it take to make available information effective? Parts of the answer have already been suggested. Available information is likely to become effective information when it is designed to attract attention, when it is vivid and personalized, when it comes from trusted sources, and so forth. The evidence of these relationships comes from general research in cognitive and social psychology and from problem-oriented experiments and program evaluation studies (Stern and Aronson, 1984:Chapter 4). More detailed answers can be expected from continued research along the same lines. Careful experiments and program evaluations can assess the effects on consumers' beliefs of different mixtures of procedures for presenting information.

The same methods could also be used to identify
indices, similar to the miles-per-gallon index for automobiles, that express a building's energy efficiency in
familiar units and so improve people's understanding of
energy efficiency in buildings. Careful field research
on suggested indices and modes of presentation would be
necessary because the most successful methods for making
full information more effective probably vary with energy
uses, types of consumer, and possibly also in other ways.

While it is possible to study the features of information and information delivery that affect people's beliefs
and action, it would be difficult to incorporate these
features into formal energy demand models. First of all,
the most effective informational efforts involve a combination of elements that probably act synergistically.
It is possible in principle to separate the effects by
experimentation or by careful analysis of a diverse enough
group of informational programs, but it would be difficult. Furthermore, the policy questions concern the
effect of information on action rather than on the beliefs
that lead to action. Thus, the simpler approach from a
policy viewpoint would be to leave out the measurement of
beliefs in the effort to identify workable means of making
information effective. This notion leads to a second
model of how information affects behavior--one that
depends less on assumptions of rational action.

Diffusion of Innovation

An alternative view of the role of information comes from
the view of social change as diffusion. In this view,
change spreads through society along social communication
networks and is subject to a range of social and economic
variables that mediate the process. When information is
fed into these communications networks, it may be transformed as it is transmitted. And through a process of
social influence, behavioral change may occur. From this
perspective, the key to understanding how energy-efficient
technologies "penetrate a market" is understanding communication and social influence, especially word-of-mouth
communication within identifiable groups or networks:
homeowners in a city, firms in an industry, municipal
governments, and so forth.

It has been suggested that innovations are adopted as
a function of their relative advantage over previous
practices, their compatibility with the adopter's values,

their apparent simplicity, the ease with which they can be adopted on a trial basis, and the observability of their outcome (Rogers with Shoemaker, 1971). Thus, the diffusion model incorporates economic cost (in the concept of relative advantage), but only after a potential adopter hears of an innovation. Information also fits in the model at that stage: it can demonstrate the advantage of adopting an innovation, make the results of adoption more readily observable, or show an adopter the results early to minimize the costs of error.

In contrast to the effective information model, the diffusion model does not hold that it is necessary for every consumer to know all the facts needed for rational action before an energy-efficient innovation will be adopted or that tangible personal benefits are the only motives for improving energy efficiency. It may be enough to get relatively full information to sources that consumers trust. Some people will assimilate all the information from these sources, but others will act on trust, without rational calculation; conceivably, trust will lead some people to do what a rational actor would do with full information.

Diffusion can also impede action. News can travel fast about the failure of an innovation to offer the advantages claimed for it or about unanticipated negative consequences of adoption. News of unsavory experiences with urea-formaldehyde foam insulation may have spread by word of mouth even before the problem was mentioned in the mass media--and the news undoubtedly affected many people who made no effort to estimate the size of the risk.

The diffusion perspective has some general validity for describing transitions to improved energy efficiency (Darley and Beniger, 1981; Stern and Aronson, 1984). It emphasizes several variables as important in determining whether information about energy efficiency will affect action: the consumer's trust in the source of information; a program's ability to get information into working communication networks; the use of vivid, simple, and personalized images or phrases of the kind people are likely to repeat; and so forth. These variables are very different from those considered important under an assumption of rational economic action, though the diffusion model also has room for variables related to costs and benefits under the concept of relative advantage.

The importance of communication and influence factors is not easily quantified, but it can be roughly estimated by field experiments that assesses the effect of extra

efforts to get information into communication networks or
otherwise to assist the diffusion process. Process evaluations of energy information programs that assess communication variables can also give an indication of the
magnitude of their effects. These research methods can
operate in the same ways we suggested for assessing the
effects of nonincentive aspects of financial incentive
programs (see Chapter 3).

While it may be possible to estimate bounds for the
importance of communication and social influence in information programs, it is not now possible to model the diffusion process itself. Only limited research has been
done to determine the processes by which energy-efficient
innovations diffuse in the society and none, to our knowledge, has used knowledge about communication in social
networks to forecast the final penetration or the rate of
adoption of particular energy-efficient technologies.
Many research questions need further attention if the
diffusion model is to be useful for those purposes. In
particular, it would be useful to know more about which
sources are trusted by different types of consumers; which
networks spread information about energy technologies
among individuals, firms, professional groups, and public
agencies; how energy information spreads through these
social networks; and what kinds of communication most
convincingly demonstrate to consumers the relative advantage of energy-efficient innovations.

For this analysis, it makes sense to use a market segmentation approach: an approach aimed at identifying
groups of people who think in the same way about a possible innovation or who trust common sources of information. This approach is valuable because different kinds
of consumers communicate in different networks and find
different kinds of information relevant and persuasive.
In addition, people use systematically different ways of
combining the elements of information that they have, and
systematically different aspects of the information affect
their decisions. To tell how people use information, it
would be useful to observe the actual decision making,
particularly among people contemplating purchase of high-
capital-cost innovations. Laboratory experiments can also
aid understanding of the kinds of information that matter
in these decisions. Survey methods may be useful for
finding out which information soures are trusted and
heeded by particular groups of consumers. Segmentation
analyses, conducted by survey or laboratory experimental
methods, can be further refined in comparison group

studies that test preliminary findings by targeting
information from existing energy information programs to
particular segments of the public.

Despite these options for improving understanding of
how diffusion affects energy-related decisions, much conceptual work and research is required before such processes could be formally modeled. It may prove useful to
integrate empirical work on the span and scope of acquaintance networks (e.g., Garevitch, 1961; Travers and
Milgram, 1969) with mathematical simulations of the processes of influence diffusion within those networks (e.g.,
Pool and Kochen, no date). Such research holds promise
in the long run for useful insight into the dynamics of
the influence of energy information.

CONCLUSIONS

People do not act always on full information about the
costs and benefits of energy-efficient investment, even
when full information is available. Analysis of energy
demand, then, depends on understanding the conditions
under which information influences action. One assumption
is that people take rational economic action based on what
they believe to be true (effective information). This
assumption is worth developing and testing. It calls for
research to assess what consumers believe about the costs
and effects of investments in energy efficiency and on the
ways these beliefs change. Surveys can determine what
people believe, but the effective information perspective
offers no guidance for understanding the conditions under
which information changes beliefs.

An alternative assumption is that innovations diffuse
through communication networks, with the quality of available information acting as one link in the process. The
diffusion perspective identifies communication and influence variables, such as trust, vividness and simplicity
of format, and face-to-face communication in informal
social networks, as important in determining whether
information is effective. Because of evidence that such
variables are important in energy choices, further
research into them is a high priority. The magnitude of
their effects can be roughly estimated for inclusion in
demand analyses by field experimentation and process
evaluation of energy information programs, but formal
models of the process by which energy information produces
behavioral change are still a long way off.

5

Behavioral Study of
Appliance Efficiency Decisions

Energy consumption is determined in part by consumers' choices about purchasing, and then about using new and replacement technology. Choices about the energy efficiency of appliances are a significant instance of technology choice for several reasons. A large part of the long-run ability of consumers to modify energy consumption patterns is embodied in the energy-using technology they choose and is fixed for the lifetime of that equipment. Two ways of affecting consumer choice, standards for appliance efficiency and incentives for selected technologies, are often considered or used as energy policy tools. And consumer information and marketing programs may also be effective tools for energy management by correcting consumers' misperceptions and encouraging purchase of more efficient appliances. Thus, wherever possible, energy policy analyses and forecasting systems should incorporate adequate descriptions of equipment technology and of choices about the purchase and use of technologies. To capture the effects of information and consumer perception that may be important in the design of policies and programs, these descriptions should incorporate a substantive account of the behavioral process by which consumers assess the alternatives and choose among technologies.

Many current policy simulation models assume that appliance technology is chosen using a criterion of life-cycle cost minimization. This assumption excludes consideration of factors that may be behaviorally important, including appliance capacity, safety, convenience, brand-name loyalty, marketing of different models, dealer characteristics, and so forth. It also ignores what may be an important interaction between efficiency and utilization decisions: consumers may use more efficient appli-

75

ances more heavily, and may anticipate this when assessing technological alternatives. Furthermore, even if the life-cycle cost criterion were an acceptable description of behavior, its realization in models now often fails to include factors that may also be important, such as consumer expectations about future fuel prices and attitudes toward risk.

In examining appliance choice, this chapter differs from Chapters 2, 3, and 4--each of which explored a class of factors that influence demand--by looking at a class of demand-related behavior that may be influenced by any of the factors discussed in the previous chapters--prices, incentives, and information. This chapter, then, is closer to the typical work of energy demand analysts: it focuses on energy demand in a particular sector of the energy economy. We first consider some of the issues raised by current models of appliance choice and identify particular areas where there are gaps in available information and in behavioral theory. We then outline the experiments and field data collection efforts that would be required for an ideal and comprehensive behavioral study of choices of appliance technology. Finally, we discuss a few practical data collection efforts that would provide useful incremental information on appliance choice.

ISSUES AND LIMITATIONS OF CURRENT ANALYSIS

There are a very limited number of major empirical bases for formal modeling of appliance fuel choices, efficiency choices, and decisions about utilization. There are a few market studies of the initial costs and technological characteristics of purchased appliances: room air conditioners (e.g., Hausman, 1979; Brownstone, 1980); heating and central air conditioning systems (e.g., Dubin and McFadden, 1984); and refrigerators (e.g., Meier and Whittier, 1983). There are also laboratory bench studies and engineering projections of the costs of appliances of various efficiencies (Hirst and Carney, 1978) and thermal models used to calculate the sizes and levels of use of heating and cooling systems in buildings with well-specified structural characteristics. In addition, there are statistical analyses of appliance purchases or holdings using household survey data--the National Interim Energy Consumption Survey and the Residential Energy Consumption Survey (RECS) of the Energy Information Adminis-

tration; the Annual Housing Survey; and national surveys
conducted by the Midwest Research Institute (MRI), the
Washington Center for Metropolitan Studies, and numerous
utilities--or state and regional cross-sectional, time-
series data. However, none of these data sources contains
sufficient information on appliance efficiencies to permit
direct study of choices about efficiency.[1] Consequently,
attention has concentrated on appliance fuel choices.

One research approach has been to assume some version
of the life-cycle cost-minimization hypothesis and to
develop estimates of the implied discount rates that
accord with observed choices (e.g., Hirst and Carney,
1978). Another approach has been to fit discrete choice
models that allow that consumers sometimes do not minimize
cost, possibly because they respond to factors other than
initial and operating costs (see McFadden, Puig, and
Kirschner, 1977; Berkovec, Hausman, and Rust, 1983; Dubin
and McFadden, 1984). The latter studies permit testing
of the life-cycle cost-minimization hypothesis, albeit in
models in which the functional forms of equations are
chosen for convenience rather than on the basis of data
or established theory. These studies generally reject the
cost-minimization hypothesis in its simpler forms. Further, the choice models they estimate meet some of the
criteria for behavioral explanation. For example, estimated parameters are reasonably stable across data sets,
and parameters estimated in some of the studies have generated accurate accounts of fuel shares among appliances
purchased in different regions and time periods (Goett and
McFadden, 1984).

But these models are still considerably short of an
adequate behavioral theory, even for fuel choice. The
models offer inadequate treatment of consumer information
and expectations, a critical limitation in light of the
policy need to predict the effects of marketing programs.
There is considerable error in measuring consumer characteristics and appliance attributes. Finally, the models
contain a large number of parameters and impose restrictions that are neither implied by behavioral theory nor
supported empirically. This does not in itself imply that
these models are useless for policy purposes. Even imperfect models may help account for physical and economic

---

[1]A partial exception is the study by Hausman (1979) of
room air conditioner choice that used MRI data on individually metered appliances in a small sample of households.

constraints and may identify variables to which consumer response is sensitive. However, policy analysts should avoid making assertions that are not supported by behavioral evidence, and improved congruence between policy models and behavioral theory should be a constant objective.

In short, existing empirical studies of appliance technology and fuel choice have important limitations, and there are virtually no studies of choices of appliance efficiency. This state of knowledge is partly due to lack of data and partly to weaknesses of theory. Questions on four topics must be addressed if analysis of appliance efficiency is to improve: (1) the analytic representation of appliances; (2) analytic representation of energy price expectations; (3) determinants of appliance use; and (4) the behavioral processes that affect consumers' decisions. Asking these questions focuses attention on the assumptions and limitations of the hypothesis of life-cycle cost-minimization as a foundation for policy modeling.

Analytic Representation of Appliances

Many questions need to be addressed in considering how appliances are represented analytically: How is the set of feasible technological alternatives defined? Is there a continuum of idealized alternatives or a finite list of brands and models? Is efficiency to be measured by manufacturer's rating, bench testing, or field testing? What dimensions of technology other than fuel and efficiency--such as capacity, noise level, size, durability, safety, convenience, and so forth--need to be included as behaviorally relevant attributes? How is the initial price of equipment defined: by list price, observed transaction price, or engineering calculation? If engineering studies are used to estimate the costs of alternative designs, what relationship holds between fabrication cost and prices? How should the prices of appliances supplied in new construction be imputed in life-cycle cost calculations: by using equipment costs to contractors or by regression analysis of building prices as a function of attributes of the equipment in the building?

Most current models are derived from the early model of Hirst and Carney (1978), which considers discrete fuel alternatives and represents the trade-off of efficiency and initial cost in terms of only three parameters. Appliance attributes such as capacity are incorporated

into the analyses only at fixed levels; the model does not
subject them to behavioral control. The estimates of
technical efficiency are determined from engineering
principles, and there is only limited validation of the
link from idealized fabrication cost to market price.
Future refinements of such models might include: representation of efficiency alternatives in terms of the
brands and models actually available to consumers; direct
measurement of purchase prices; representation of efficiencies as rated by manufacturers; field testing of realized efficiencies; measurement of attributes other than
purchase price and efficiency; data collection on the
degree of discounting by distributors from list prices,
particularly in new construction; and measurement of the
"perceived value" of equipment alternatives in new construction by hedonic regression.[2]

Analytic Representation of Energy Price Expectations

In considering energy price expectations, two major questions stand out: How are future energy prices, as anticipated by consumers, to be estimated? Which assumption is
most justified for behavioral models of expectations:
static prices at current levels, perfectly anticipated
prices, or prices changing at historical trend rates?

Most current studies assume a static expectation that
current real fuel prices will continue to prevail in the
future. Recent econometric investigations of expectation
formation (primarily of financial markets, for which considerable data are available to market participants and
to observers) suggest that consumers are less naive and
use information to form "models" of future market behavior
(Eichenbaum and Hansen, 1983; Mishkin, 1983). The limited
evidence on energy markets is consistent with an
adaptative-expectations hypothesis for gasoline prices
(see Chapter 2), but this conclusion may not be applicable
to electricity and natural gas markets, in which con-

---

[2] In this procedure, detailed data on actual equipment
choices or laboratory simulations of equipment choices
are used to regress prices paid for appliances on a
number of appliance characteristics measured or simulated
in the study. The resulting regression equation represents the "perceived value" of each of the characteristics.

sumers' decisions are less routine and information is less systematic. Still, the importance of expectations for policy affecting appliance purchases makes the behavioral process of forming expectations a critical topic for study.

Determinants of Appliance Use

In studying appliance use analysts should address at least two major questions: How is anticipated future use of appliances estimated? If the level of use is not assumed to be fixed and beyond behavioral control, how does one incorporate feedback from appliance operating costs into level of use in estimating life-cycle cost?

Current models using the criterion of life-cycle cost-minimization assume that, at least from the standpoint of a purchaser at the time of an appliance choice, the intensity of appliance use is fixed and beyond behavioral control. While some of the models allow for changes in use once appliances are in place, this approach implies complete lack of planning by consumers. The possibility of feedback from anticipated operating cost to utilization to choice of technology poses a fundamental problem for the life-cycle cost-minimization hypothesis, since it implies that consumers may trade initial costs against the benefits of greater future utilization: for example, by buying an energy-efficient air conditioner, a household may be able to afford to keep cool more of the time. Assessing this possibility requires study of the joint choice of technology and utilization, with a theory of choice that allows trade-offs and with survey data that encompass both purchases and subsequent appliance use.

Consumers' Decision Processes

Analysts should address several questions on decision processes: How are future costs to be discounted under the cost-minimization hypothesis? How are appliance durability and consumers' replacement strategies taken into account? Are there perceptual elements that lead to different discount factors in fuel choice and in efficiency choice? What is the impact of uncertainty about appliance performance or about resale market value? What is the impact of credit constraints?

A number of studies have estimated discount rates by assuming the weights given to initial and operating costs in discrete choice models of appliance fuel choice. These studies have usually ignored the effects of expectations about fuel prices, interest rates, and equipment life and of feedback from utilization. The empirical results are quite erratic, reflecting these factors as well as problems in accurately measuring initial and operating costs and limitations of the life-cycle cost-minimization hypothesis as a behavioral model.

Refinements require a recognition that the life-cycle cost-minimization criterion must be modified under at least three conditions: (1) when the alternatives vary not only in energy efficiency but in service attributes, such as capacity and convenience; (2) when there is feedback from anticipated levels of appliance use; (3) when consumer behavior is likely to be more complex than the criterion suggests. Some of the objectives of a behavioral theory may be achievable by adopting a psychophysical version of the traditional economic model of preference maximization, in which consumer behavior is assumed to vary around optimal choice according to some mathematical distribution. This is a relatively mechanical approach to accounting for deviations from life-cycle cost minimization. It is easily adapted to the task of quantitative simulation (see Thurstone, 1927; Tversky, 1972; Luce, 1977; McFadden, 1981), but it does not provide a useful framework for studying the relationships of information processing to choice or the social aspects of the transmission or use of information.

A second approach is to conduct experimental or marketing studies of specific behavioral phenomena. Two examples are a controlled field experiment to test the effects of alternative cable television advertising "treatments" on consumer awareness of appliance costs and the use of focused group discussions to investigate the structure of consumers' information networks. Such studies can potentially provide information critical to the design of marketing programs for energy-efficient appliances. However, they are not well suited to developing general purpose quantitative policy simulations.

THE IDEAL DATA FOR BEHAVIORAL STUDY
OF APPLIANCE CHOICES

A comprehensive behavioral study of choices affecting appliance efficiency would concurrently examine the

effects on appliance choice of fuel, efficiency, expected
utilization levels, and a range of features of appliances
and their marketing. It would also address the efforts
of government and utility companies to influence consumer
choice. This comprehensiveness imposes requirements on
both data collection and on the method of analysis. The
ideal data base must contain descriptions of all appliance
alternatives, chosen and unchosen, as well as of consumers' actual choices and their subsequent intensities
of use of the appliances. The efficiencies of alternative
appliances should be measured both as perceived by consumers (e.g., manufacturers' labels, the content of advertising, and attitudes about the appliances) and in terms
of field performance. The initial prices of the appliances should be measured as actual transaction costs.
Appliance use should be observed under a sufficient variety of conditions to reliably assess behavioral response
and that information should be used in describing any
feedback from utilization to appliance choice.

To carry out the collection of such data in a controlled setting, with availability and attributes of
appliances, prices, and other factors under experimental
control, would be a formidable undertaking. Less defini-
tive, but nevertheless informative studies, could be carried out with field data from natural experiments. Collection of such data by the U. S. Department of Energy
would require a different orientation from that of annual
surveys such as RECS that provide census-like data on
energy consumption. Such studies would necessarily be
smaller and more specialized, with intensive data collection. Most of the elements of such surveys are already
in use by various electric utilities, which collect
appliance serial numbers, develop panel surveys, and
attach meters to individual appliances. The U. S. Bureau
of Labor Statistics routinely collects data on appliance
transaction prices and uses hedonic regression techniques
to relate price to equipment attributes. Without a
national framework for simultaneous collection of such
data, however, it is unlikely that enough standardization
and comprehensiveness will be achieved to give good answers to questions about how consumers make choices about
the energy efficiency of purchased appliances.

Any general analysis of these questions based on field
surveys will certainly leave some key questions unan-
swered. Specialized experiments and marketing studies
may shed light on specific behavioral phenomena that are
critical to particular policy problems and may help make

sense of the relationships observed in field data. For example, research by cognitive psychologists shows that people's choices can depend on the way in which the alternatives are presented to them (Tversky and Kahneman, 1981). Small-scale experiments could be conducted to explore the implications of this phenomenon for the marketing of energy efficiency in appliances.

With regard to methods of analysis, there will always be a conflict between researchers who hesitate to extrapolate from weak empirical evidence and policy analysts who seek the least inadequate method of extrapolating as far as is required by an immediate question or problem. Given the current state of behavioral theory, the limits to experimentation with human subjects, and the problems of understanding complex behavior by use of survey data, any practical policy tool will almost inevitably severely compromise scientific standards. Consequently, policy modeling should be a continuing process of scientific attack, invalidation, and improvement of interim simulation tools. With specific reference to the problem of analyzing appliance efficiency standards, an initial goal might be to carry out the data collection and analysis necessary to describe efficiency choice at the level of empirical precision that has already been achieved for fuel choice, while at the same time developing the experiments and behavioral knowledge necessary to refine the descriptions of both fuel choice and efficiency choice.

PRACTICAL ALTERNATIVES FOR DATA COLLECTION

Most of the quantitative policy models currently in use at the U. S. Department of Energy and elsewhere are based on the assumption that equipment efficiency choices are governed by life-cycle cost-minimization criteria or on some variant that admits exceptions in which additional aspects, such as capacity, may influence choice. The most narrowly defined cost-minimization models would be rejected by most social scientists as behaviorally unrealistic, particularly because of their insensitivity to such factors as consumer information, which may be important foci of policy. As a result, most existing models would be judged inaccurate for forecasting. Such models, if modified to make them consistent with classical psychophysical and economic "laws" and fitted to survey data, may offer reasonably accurate bases for simulations of policies that operate primarily through economic vari-

ables, but even these models still have to be validated
for such applications.

One task for those people who challenge current models
is to propose improved alternatives that are practical
enough to be adopted by policy makers and planners. A
recommendation to scrap the current generation of models
may be sound, but is unlikely to be followed (as indicated
in Chapter 1). What may be easier is to move away from
the current models toward directed experiments and field
studies when possible and to continue to try to improve
the behavioral foundations of current simulations. One
way to implement such an approach would be for the U.S.
Department of Energy (DOE) to cooperate with the private
sector whenever feasible to promote standardization and
sharing of data sources and to further develop data collection methods, such as automatic metering of individual
appliances. Such coordination might involve industry-wide
groups such as the Electric Power Research Institute and
the Gas Research Institute. The DOE could also cooperate
with other federal agencies in data collection. It would
be helpful, for example, if DOE promoted joint collection
with the Bureau of Labor Statistics (BLS) of data on the
transaction prices of appliances and encouraged BLS to
refine and publish hedonic regressions accounting for
housing prices as a function of building attributes. The
DOE should also review its use of large-scale policy-
simulation models, adopt an ongoing program of upgrading
and evaluation, and redirect policy analysis of appliance
choice toward using specialized experiments and surveys
as much as possible.

The central element in a program of data collection on
appliance efficiency would be a household survey that
covers appliance choice, including fuel, efficiency, and
other features, and that subsequently meters the chosen
appliance. The Electric Power Research Institute is
developing the technology for the last of these, appliance
metering. A random survey sample could be drawn from the
population with further selection of a sample of recent
appliance purchasers. Alternatively, information could
be collected retrospectively on old appliances, although
this possibility is limited by the ability to obtain
retrospective data on price and rated efficiency. Another
design would measure appliance use among a panel of purchasers. This last approach would require careful statistical analysis and would lack useful observations on
utilization before appliance purchase. Measurement of
appliance efficiencies and ratings could be carried out

in a manner similar to Consumer Reports bench studies or
EPA fuel efficiency measurements for automobiles; some
utilities are now collecting such data with the coopera-
tion of manufacturers, using appliance serial numbers.
By cooperating with the private sector, DOE may be able
to improve its own capacity to forecast the quantitative
effects of energy policies on appliance efficiencies.

6

Conclusions and Recommendations

The previous chapters examine several factors that may be important influences on energy demand but that have not been addressed extensively in existing formal models. Some of these, such as consumer mistrust of information, informal social influence, and the marketing of financial incentive programs, are rarely highlighted by modeling efforts and may be difficult to assimilate into existing models. Such factors, it might be said, are blind spots of existing models. Other factors, such as qualitative distinctions among types of financial incentives, are not emphasized by the theoretical frameworks on which most models are based, but are likely to emerge in modeling efforts and could be assimilated into models without very much difficulty. Still other factors, such as the distinction between appliance list prices and transaction prices and the effects of price changes as distinct from price levels, are significant in economic theory and could be incorporated readily into existing models, even though they have not been in the past. Our analysis shows, in sum, that existing demand models describe the behavioral environment of energy demand only incompletely. Some of the gaps can be filled if models are built from more complete and detailed data, but modeling efforts are likely to overlook systematically some important features of the environment of energy use.

Because some important gaps in demand analysis seem unlikely to be filled through modeling efforts, we judge it unrealistic to try to build a single comprehensive analytic framework to answer policy makers' and researchers' many questions about energy demand. As an alternative to seeking such a framework, learning should proceed by developing a portfolio of analytic approaches--some general, others focused on particular policy questions.

With diverse methods of analysis, hypotheses will be generated and tested that any single method might overlook.
In addition, one method can sometimes provide a valuable
check on the results of another.

This chapter presents the panel's conclusions and
recommendations regarding the use of formal models and
problem-oriented research in energy demand analysis and
regarding the needs for data to inform that analysis. We
also discuss an approach to using different analytical
approaches in concert in order to improve the quality of
energy demand analysis.

THE ROLE OF FORMAL MODELS

Policy analysts sometimes express two erroneous opinions
concerning formal energy models: one is that a policy
question can be answered simply because it is represented
in a model; the other is the obverse--that a question
cannot be answered because no model exists to answer it.
Both opinions equate energy policy analysis with formal
modeling. They reflect an overreliance on formal models
that is not justified by the validity of existing models
and that is not necessary given the availability of other
analytic techniques.

Formal models do have considerable appeal as a means
of energy policy analysis. They are broad, multipurpose
tools that can address a wide range of policy questions
and call attention to unanticipated effects of policies
on other parts of the energy or economic system. They can
give the sort of quantitative responses decision makers
want to their questions, and they can often do this
quickly. And when correctly formulated, models can provide necessary checks of consistency with physical and
economic constraints that might otherwise be overlooked
in a policy analysis. Compared with methods that involve
gathering new data, models can save both money and time.
They can also evolve, along with the questions that face
policy makers. From a policy maker's viewpoint, models
are familiar tools once they have been used, so it is easy
to continue to rely on them, sometimes even when they are
outdated. In addition, the apparent objectivity of computerized analysis is impressive to some decision makers.

But models have many limitations. As the previous
chapters have demonstrated, there is no behavioral knowledge to support assumptions about the values of parameters
and about the functional forms of the equations used to
represent behavioral relationships. Many models do not

treat their internal uncertainty with sufficient explicitness. And important variables are often excluded from models, sometimes for lack of data and sometimes because modelers' conceptual frameworks do not include them.

Another limitation of models that are used for policy analysis and forecasting is that they are usually validated only by matching them to past events, often with numerous post hoc adjustments. This procedure does not justify confidence in a model's ability to predict the future: newer versions of a model may be as likely to need readjustment as the older ones.[1]

The documentation, validation, and maintenance of models have generally been given insufficient attention. A model must first be documented: complete records must be made of the model's contents, its assumptions, and the sources of the parameter estimates and the chosen functional forms of its equations. Only with full documentation can a model's behavioral assumptions be identified for testing. Validation is also essential, and not only when a model is built. In our view, it is important to validate models by testing them regularly against empirical data. This means, for instance, making before-the-fact predictions with a model and comparing them with actual outcomes or comparing a model's results with the results of problem-oriented studies. Such empirical testing may be the only way to build credibility for models, in light of the fact that many models undergo almost continuous revision in their structure, input, and output. This ongoing validation, combined with a commitment to updating the documentation of a model as it evolves, constitutes maintenance--a much neglected part of modeling. When an organization buys expensive capital equipment it usually commits itself to a budget for maintenance. But energy models, which are not as reliable as most equipment, are often expected to work well without maintenance.

The size and complexity of some energy models makes documentation, validation, and maintenance particularly

---

[1]This shortcoming of models is due not to the nature of modeling but to the frequent practice of inferring regularities in human behavior from the evidence of past correlations. Other research methods, particularly survey methods and exploratory data analysis, have the same shortcoming when their findings are used uncritically to make projections.

difficult. Although larger models are being better able
to include feedbacks among parts of the energy system, the
burden of documentation and validation increases, sometimes geometrically, with the number of relationships
represented. Models that include large numbers of parameters compared to the volume of data they explain are
particularly suspect. Also suspect are large models,
including many of the system dynamics variety, whose
results are sensitive to the effects of variables whose
values and relationships are merely postulated. Given the
state of the art, we conclude that more knowledge can be
gained by improving the quality of models than by increasing their size. With constant resources, modeling can be
improved, on the whole, by sacrificing some comprehensiveness in order to gain quality. This ongoing validation,
combined with a commitment to updating the documentation
of a model as it evolves, constitutes maintenance--a much
neglected part of modeling.

In addition to the above substantive problems, the
process of funding for models gives cause for skepticism.
When quick answers are in greater demand than documentation and validation, model builders are under pressure to
sacrifice quality control. Poorly validated models can
be expected to be used more often and better--and therefore more expensive-- models will fail to command the
support their higher quality deserves.

As a result of all the above factors, when any existing
energy demand model gives an answer to a policy question,
that answer is to a large extent taken on faith. Despite
these limitations, models remain popular with policy
analysts--so popular that they are sometimes overused or
misused. Sometimes models are used to answer factual
questions that could be answered almost as easily and
much more accurately by other methods. When an oil shortage threatens, for example, it makes more sense to find
out how much consumers are adding to their inventories by
surveying a sample of consumers than by estimating behavior from a model. Sometimes models are used to answer
policy questions that they are not equipped to address.
For instance, most models have difficulty representing
efforts to improve information. To estimate the effect
of energy-efficiency labels for appliances, a modeler
might postulate an effect of the labels on consumers'
discount rates and use a model to estimate that effect on
purchase behavior. It would make more sense to conduct a
field experiment that actually tested the effects of
labels.

Models may be quick and inexpensive relative to alternative research methods, but there is no such thing as good, cheap energy policy analysis. If policy analysts are to offer knowledge rather than mere answers, the empirical basis of their analyses must be strengthened. We believe this can be done by making some changes in the way models are developed and by drawing more on knowledge gained by other methodologies. We offer seven conclusions and recommendations about formal energy models and their use.

1. Policy makers should maintain a healthy skepticism about the outputs of formal energy demand models. We do not assert that judgment is necessarily better than existing models. Rather, the point is that models, like judgments, should not be accepted without corroborating empirical evidence. The support of a second model is much less convincing evidence than the support of a field experiment, a good evaluation study, or even a well-conducted survey.

2. The current system dynamics models in use at the U.S. Department of Energy should not be relied upon as heavily as they are for forecasts of energy demand. Forecasts from those models are too dependent on postulated relationships and on judgmental elements incorporated in them to make them consistent with expectations.

3. Resources allocated to modeling should be shifted to ensure adequate documentation, validation, and maintenance.

4. Within the modeling community, more attention should be paid to building models that are better tested and maintained. These efforts are necessary to make demand models more credible.

5. For testing purposes, some versions of some models should be "frozen," archived, and then used from time to time without judgmental readjustments to make forecasts and policy analyses that are then tested against new data and against the findings of studies that use other research methods. This step should be considered an essential part of the validation process, and is conceptually separate from the normal process of using new data to revise and update models. With this step, the modeling

community can build a track record on the basis of which formal models can be judged.

6. Innovation in modeling should be directed toward decreasing the dependence of model outputs on assumptions about parameters and the functional forms of equations.

7. Sensitivity testing of models should be used to generate hypotheses for empirical research, and resources used for validating models should be devoted in part to carrying out this research. When a model's output is highly sensitive to a parameter whose value is not well established, empirical research should be done to establish the value.

As these conclusions and recommendations make clear, we believe much more emphasis should be given to building the empirical base for energy demand analysis than to further elaboration of formal models based on inadequate data. Other research methods are required to build this empirical base.

THE ROLE OF PROBLEM-ORIENTED RESEARCH

Five types of problem-oriented research are surveys, analyses of existing data, natural experiments, controlled experiments, and evaluation research.

Surveys

National general-purpose surveys can provide invaluable data for problem-oriented research. Because their primary role in energy demand analysis has been to gather the multivariate time-series data essential for much policy analysis, including formal demand modeling, our conclusions and recommendations for these surveys appear in the next section on data collection.

Specialized surveys have been responsible for most of the detailed analytical work on the effects of consumer knowledge, attitudes, and beliefs on energy use (e.g., Kempton, Harris, Keith, and Weihl, 1982; Stern, Black, and Elworth, 1982b). Specialized surveys are especially useful for explaining phenomena that appear inconsistent in terms of the variables represented in models. For example, surveys can be used to understand why conserva-

tion programs that offer the same financial incentives
vary so widely in their levels of consumer acceptance (see
Chapter 3; Berry, 1982). Surveys that explore public
responses to the marketing and implementation of conservation programs have shown that consumer protection and
convenience are two nonfinancial variables that affect
consumer response to financial incentives (e.g., Stern,
Black, and Elworth, 1981).

But as we have mentioned, surveys also have limitations. One is the unreliability of self-reports of some
variables, such as attitudes. There is evidence both of
reliability and unreliability in responses to energy surveys (e.g., Beck, Doctors, and Hammond, 1980; Geller,
1981). Two reasonable but unproven hypotheses are that
self-reports of major investments are more reliable than
self-reports of changes in habits and that reports of past
action are more reliable than reports of future action.
There is reason to question the worth of self-reports
about planned energy-saving actions. Although reported
intentions to act are often good predictors of behavior
(Ajzen and Fishbein, 1977), the relationship depends,
among other things, on the absence of constraints on
action. For expensive investments in energy efficiency
that involve many steps before completion, behavioral
intentions would seem a questionable predictor. We have
also noted that inferences drawn from even the most
accurate self-reports may not be accurate because of
errors in analysts' assumptions relating energy-saving
actions to subsequent energy use. And when a policy
innovation is being considered, people's predictions of
how they will respond to hypothetical situations are not
as good a source of information as actual observation.
For such situations, small-scale experiments and program
evaluation studies can give more useful information, even
if their generalizability is unknown. The conjunction of
several small-scale behavioral studies can give more confidence in the conclusions about a new policy than the
best-designed national survey of people's intentions.

Analysis of Existing Data

Analysis of existing data can still improve understanding
of energy demand. Many interesting data sets collected
by government agencies go unanalyzed, in whole or in part.
For example, the data from the Residential Energy Consumption Survey (RECS) have been only partly analyzed.

Inadequate information about individual respondents is
cited by some researchers as one reason for the lack of
detailed analysis, but lack of funding for data analysis
is a more serious deterrent. More analysis could also be
done on utility companies' data on residential and commercial energy use to give a more solid empirical basis
to studies of energy demand. Limited funding for research
and the narrow focus given to research questions have
limited what could be learned; noncomparability of data
across utility companies has made analysis difficult; and
access to the data has been a major problem. Concerns
about customer privacy and about possible use of data in
adversary proceedings make many utility companies unwilling to give researchers access to their files. Regulatory
agencies have sometimes forced the release of data when
they believe release will serve a public purpose, but as
a rule, utility data are not readily available to
researchers.

Existing data can be studied in various ways to learn
about energy demand and to generate hypotheses. Several
techniques of exploratory data analysis (Breiman, Friedman, Olshen, and Stone, 1984; Donoho, Huber, and Thoma,
1981; Fisherkeller, Friedman, and Tukey, 1974; Friedman
and Tukey, 1974; Huber, 1981) can be used to examine data
sets for regularities and to generate hypotheses to be
tested on future data or with additional research. These
methods rely heavily on informal graphic techniques and
deemphasize formal statistical models or tests of
hypotheses.

8. We recommend that some of the resources devoted to
energy demand analysis be redirected toward exploratory
analysis of existing data.

Disaggregated data should be systematically collected
on energy use in the commercial and industrial sectors of
the economy and on energy prices and equipment stocks.
Specialized surveys relating measured energy use and
observed investments in energy efficiency to demographic,
institutional, and attitudinal factors are also much
needed.

Natural Experiments

Natural experiments can produce a wealth of data that
should be analyzed systematically. Ongoing data collec-

tion efforts would make this more possible. Analysis of the records of utility companies would allow additional studies. Natural experiments would teach more if a capability were developed to move researchers quickly into the field to study natural experiments in energy demand. Surveys of consumer response to changing utility rates or the recent decrease in inflation rates would have been a way to learn from one class of natural experiments (see Chapter 2); studies of initial responses to the threat of an oil supply cutoff could be a vehicle for learning from another class of natural experiments.

   9. We recommend that some of the resources available for energy demand analysis be made available on short notice for field studies of natural experiments that occur when there are rapid changes in the energy environment.

Controlled Experiments

Controlled experiments are an especially valuable tool for assessing the effects of interventions that are non-financial in character and for which existing models are particularly inadequate. For example, psychologists have conducted many field experiments on the effects of energy-use feedback (reviewed by Geller, Winett, and Everett, 1982) and smaller numbers of field experiments to assess the effect of nonfinancial factors such as personal commitment (e.g., Pallak, Cook, and Sullivan, 1980), self-monitoring of energy use (e.g., Becker, 1978), and the presentation of energy conservation as a way to save money versus as a way to avoid losing money (Yates, 1982). Financial incentive programs are also appropriate subjects for field experiments (see Chapter 3), both because they have important nonfinancial features and because consumer responses to the incentives themselves are not well understood.

   Experimental techniques offer great benefits for policy analysis of conservation programs: controlled field experimentation should be the method of choice for evaluating promising innovations in the implementation of such programs. Conservation programs are complex and contain important elements of promotion and implementation that cannot easily be expressed or analyzed in models. For example, results from Residential Conservation Service (RCS) programs have varied greatly across the utilities that run them, leading to controversy about whether the

national RCS program is worthwhile, and conflicting judgments have been offered to policy makers on the basis of
very weak evidence. But many of the likely sources of
variation could easily be the subject of experimentation
at low cost. A utility company could randomly assign some
of its customers to receive telephone marketing efforts
or to be contacted as a follow-up to energy audits, or to
receive lower-cost audit procedures as controlled alternatives to the procedures the utilities now use. Despite
the fact that strong inferences could be drawn from such
experiments, conservation programs are almost universally
designed and implemented without the controls necessary
for identifying low-cost means to improve their chances
of success.

10. <u>We recommend that controlled field experimentation
be used whenever possible to evaluate promising innovations in policy affecting energy demand.</u>

As we have mentioned, laboratory experiments also are
appropriate analytic tools in some circumstances. They
are particularly useful in efforts to design energy
information so that consumers will notice and understand
it (see Chapter 4). It is often feasible to experiment
in a laboratory setting with alternative choices about
what information to include, what metrics to use to sum-
marize information, and how to design appliance labels,
automobile fuel economy guides, utility bill inserts, and
so forth. The laboratory approach is much cheaper than
field experimentation and can be used to screen out
alternatives that would almost certainly fail in field
trials.

Evaluation Research

Evaluation research can, at least in principle, allow
analysts to learn from what may be the greatest untapped
source on information about energy demand--the thousands
of energy programs and policies that have been tried during the last decade. The knowledge that could be gained
has great practical value because the success or failure
of a conservation program is probably due to more than the
sum of the specified features it offers; thus, it is not
enough to build a program from single features that have
proved effective--even in well-controlled experiments.
The experience gained in past programs and policies, if

it can be interpreted, can help identify and possibly
harness forces that may be more important than many of
those usually considered in formal energy analyses. The
best example is the fact that consumer use of incentives
for conservation can vary by two orders of magnitude among
programs offering the same financial incentive (see Chapter 3). This finding presents a riddle for analysts if
they define the programs simply in terms of the financial
value of the inventives they offer. The riddle can probably be solved only by carefully examining the ways the
different programs are implemented. Such process evaluations, which emphasize qualitative research methods based
on close observation and interviews of program staff and
clients, can offer the needed insight. Outcome evaluations, which can use many of the research methods discussed in this section, can offer quantitative estimates
of program effects. Careful comparisons of outcome
studies can also provide estimates of how much difference
process factors make.

Although much can be learned from thorough process and
outcome evaluation of the experiences of energy programs,
we wish to reemphasize that the most reliable information
comes from explicitly treating programs and policies as
experiments from their beginning. Such an approach
requires the creation of a suitable comparison group,
randomly assigned if possible, and careful measurement of
effects in all groups (fuller accounts of issues in evaluation research design can be found in texts such as Cook
and Campbell, 1979). Experimental research methods do not
imply, we repeat, rigid constriction of a program's operation for the sake of some notion of scientific rigor.
When controlled experiments are not feasible, some quasi-
experimental research designs retain many of the advantages of controlled experiments. Whatever the type of
research design, however, more can be learned from the
experience of a program if an evaluation plan is developed
as a program is developed; an evaluation plan tacked on
after a program has been operated inevitably produces
weaker research because of the inability to measure pre-
program conditions and because important questions must
be answered from memory or by reference to incomplete
archives rather than by observation.

11. <u>Resources devoted to energy demand analysis should</u>
<u>be shifted to favor collection and analysis of empirical</u>

data over further elaboration of models that are poorly
supported empirically.

12. Additional efforts should be made to identify and
quantify important variables that are now omitted from
formal energy models. The most obvious example is the set
of marketing and implementation variables that appear to
dwarf the effect of financial incentives in energy conservation incentive programs. Evaluation research appears
to be the best method for identifying the relevant variables; evaluation research or field experimentation might
be useful for estimating their size.

13. Additional analytic effort should be made to
incorporate key nonfinancial variables into the process
of demand analysis. The effects of marketing and management in energy information programs or of consumer mistrust may be interpreted as changes in discount rate,
changes in lag coefficient, or in other ways. It will
prove valuable, however, not only to quantify the important nonfinancial factors in energy demand but to improve
their conceptualization.

14. The federal government should establish a fund
for basic research on decision making relevant to energy
efficiency, with grant awards recommended by an outside
peer review panel. Such research should include studies
of nonfinancial influences on energy demand and studies
with only indirect implications for existing government-
supported energy programs.

THE ROLE OF DATA COLLECTION

In efforts to model energy demand, data on energy use and
on factors that influence it have too often been imputed
rather than measured. Energy use is often calculated from
data on production, stocks, and imports and then allocated
to end uses, sectors of the economy, and geographic
regions. Data on energy use by energy-efficient technology are often estimated from engineering models rather
than measured in actual operation. And the nature of
consumers' and manufacturers' decisions, program implementation, and other social processes is most often
assumed (or ignored). Insufficient knowledge exists to
justify relying on imputations or presumptions rather than
measured data. Prudence dictates building some national

estimates from disaggregate measurements and surveys, more direct methods that can act as a check on procedures of imputation. It makes sense for such measurement efforts to emphasize major energy uses (e.g., gasoline for automobiles); politically sensitive uses (e.g., home heating, which especially concerns low-income consumers and their advocates); uses for which major fuel switching is possible (e.g., industrial process heat); and uses about which little is known, such as energy use in commercial and public buildings.

The best current example of national data collection on energy demand is the Residential Energy Consumption Survey (RECS) of the Energy Information Administration (EIA), a detailed longitudinal survey of a rotating panel of households that has been a particularly important source of knowledge for demand analysts. Careful thought has gone into the construction of the RECS questionnaires, which have served as a model for some other surveys and could be used more in research by state and local governments and by utility companies.

For several reasons, however, national surveys have not achieved their potential. For example, the initial plan for EIA to survey energy use in nonresidential sectors of the economy has not been followed. Understanding energy demand in the industrial and commercial sectors--the bulk of national energy demand--is obviously critical for national demand analysis, yet the EIA survey of industrial energy use was abruptly discontinued in 1981, and a planned new survey has not yet appeared. The survey of nonresidential buildings has been a sporadic effort and deserves more support. And EIA's data on transportation are restricted to the residential sector. These weaknesses in EIA's surveys should be corrected.

15. <u>Serious and continuing support should be given to EIA surveys that address all major energy-using sectors of the economy, that use a panel design, and that are conducted by experienced and competent data collection organizations</u>.

16. <u>The industrial energy-use survey of the Energy Information Administration should be reinstated</u> to gain essential data on a major segment of national energy demand.

Technical problems have made it difficult for some researchers seeking to use the RECS public data base.

Data tapes are not available for up to 2 years after the
data are collected.[2] More important, details at the
individual level, which analysts often need for micro-
analysis, are not available from RECS because of concerns
about privacy, disclosure, and informed consent. In particular, these concerns have resulted in limiting information available about the specific location of respondents' homes. Without this information, however,
researchers cannot take advantage of information available
from other sources on such factors as prevailing wind
speed and direction, differences in utility rate structures, the exposure of households to local or state conservation programs, or local consumer price indices.
Information at the level of three digits of a zip code
would allow analysts to assess the effects of local variables more adequately than they now can. The privacy
problem might be solved by requesting respondents to
release more detailed information to investigators, by
relying on smaller surveys in which participants volunteer
to release the information needed to answer particular
questions, or by allowing the data collection organization
to merge a researcher's data set with the RECS data for a
subscription fee. EIA has occasionally merged data sets
or done additional data analyses on the request of and
with funding from other federal agencies.[3]

17. The Department of Energy should, wherever feasible,
cooperate with other federal agencies and the private
sector in data collection.

RECS has also failed to include enough detail to be
useful to certain specialized groups of researchers. For
example, it has not assessed the importance of energy
efficiency and other factors in appliance purchases. It
has also done little to assess motivational and social-
psychological factors in energy demand. Of course, there
are limits to how much a survey can include, and some

---

[2]The delay is due at least in part to the operational
difficulty of collecting and checking data from disparate
sources. For example, RECS must collect data from households and subsequently from energy suppliers. It can take
six months or more simply to collect energy use data from
fuel oil dealers.
[3]Information from L. Carlson, Energy Information
Administration.

potentially important questions will always be left out.
We do not offer proposals to restructure the RECS survey,
but we do believe it should be improved. And all of EIA's
surveys should be designed to obtain the best and most
useful data for research and policy analysis.

18. A formal advisory board of energy demand
researchers should advise the Energy Information Administration on the contents of its surveys.

19. Continuity is a high priority in data collection.
Surveys of energy consumers should repeat items over time
and use a panel or rotating panel design.

We also wish to emphasize the occasional need to gather
representative national data on energy issues on short
notice or at relatively little expense. For example, EIA
conducted a survey in fall 1979 of the oil-heated households in its national sample to see if people were having
trouble obtaining heating oil in the wake of the oil
shortage of that year (Energy Information Administration,
1979). The existence of a well-chosen representative
sample for which baseline data were available made it
possible to conduct a survey on short notice from which
meaningful conclusions could be drawn, and we believe such
samples should be maintained.

20. A large national panel for which past data exist,
such as the RECS respondents, should be made available
for subsampling so quick telephone surveys can be used to
help answer immediate policy questions. Such a subsample
might be made available to independent researchers who
could insert questions on a subscription basis.[4]

---

[4]We have not addressed legal questions that may arise
from selling subscription access to respondents to a
federally sponsored survey, particularly to profit-making
organizations. The point is not that the RECS survey
should necessarily be the vehicle for collecting the
data, but that some preexisting national survey would be
valuable as background for more focused survey efforts by
public or private organizations. In the residential
sector, RECS is the best such survey in existence.

USING VARIOUS RESEARCH METHODS IN CONCERT

Energy policy analysis and related data collection tend
to be closely responsive to policy questions: current
policy issues drive the development of models, and the
requirements of models determine data collection efforts.
Immediate policy questions and formal modelling tend to
dominate the research enterprise to the neglect of other
methods and of more basic research. The demand for
answers to today's questions today has diverted resources
from the more basic task: building a knowledge base for
answering policy questions more accurately. Instead,
emphasis has been given to elaborating formal models even
when their assumptions are poorly tested, the necessary
data are lacking, and important variables are not included
in them.[5]

We have noted the important place of formal models in
energy analysis, and we believe that because of their
great value for forecasting and for identifying effects
of policies on disparate parts of the energy system,
improving their behavioral foundation is a high priority.
Models can also be useful for more narrowly focused policy
analyses, though they have been overused in relation to
other methods. In this context, models are most appropriate for anticipating effects of interventions that are
quantitative and that operate by processes that are well
understood or that have been successfully modeled in past
similar situations. In the more typical case, however,
when the path of implementation is less straightforward
(e.g., energy conservation tax credits, regulations,
informational efforts), existing models are less useful.
They have even less value for analyzing policies that are
qualitative in nature, or that obviously involve institutional, organizational, or psychological elements (e.g.,
residential conservation programs). For such analyses,

---

[5]Models are most often constructed by engineers,
operations researchers, and economists, with little
consultation with researchers in other disciplines. This
lack of breadth is one reason there has been so little
effort to model such variables as incomplete information,
communication processes, marketing of programs, and
decision under uncertainty. Data on these variables are
hard to get, but the effort has not seriously been made
and the variables tend, as a result, to drop out of
consideration in policy analyses based on formal models.

problem-oriented studies are more likely to offer useful information to policy makers.

In short, although good energy models are desired and needed, existing formal models are not yet up to the tasks for which they are used. Better analysis requires a serious research and data collection effort driven not only by immediate policy concerns but by a desire to improve understanding of energy use and general theories of consumer behavior.

Such an effort implies changes in the use of formal models and other research methods. Formal models need input from other research methods, which are especially useful for supplying empirical tests of modeling assumptions and predictions. Researchers should use the various research methods in a complementary fashion, using each to answer the kinds of questions for which it is best suited and, when more than one method is appropriate, using each as a check on the others. As a general strategy, we advocate a combination of research methods as the best way to advance understanding of energy demand.[6]

Problem-oriented research can combine with models in various ways. The results of some problem-oriented studies raise questions about which variables are most important to consider in formal demand analysis. For example, the data from evaluation studies on the wide disparity of response to a constant financial incentive suggests that something about the implementation of incentive programs (not now represented in models) may be more important than the monetary value of the incentive (a

---

[6]Our discussion of the character of multimethod research on energy demand is not meant to minimize the real institutional barriers to making this a normal part of policy analysis. The people who construct formal models and those who use other research methods often come from different disciplinary backgrounds, belong to different professional associations, and communicate little with each other. And in policy-making organizations, there is often a similar split between units that do modeling and units that do other research, for example, program evaluation. There are some signs of improved communication, including some interdisciplinary conferences on energy demand issues and the existence of the present study--but the problems institutionalizing a multimethod approach still loom large.

major focus of modeling efforts). Problem-oriented
studies are the only available way to estimate the effects
of variables that are not now represented in models.
Evaluation research on the implementation of conservation
programs is one example; another is research on improving
the quality of information available to energy users with
feedback (Chapter 4).

Problem-oriented research methods, in concert, can
provide empirical help in estimating the parameters of
models. For example, to estimate the effect of appliance
labels that offer information on energy efficiency, small-
scale laboratory experiments might first be used to determine what information is effective on labels and what
presentation formats people consider useful. Field
experiments in which labels are used in some locations and
not in others would be the best way to get a realistic
estimate of how much difference the best available labels
make. Surveys of appliance purchasers can produce empirically based estimates and act as a check on the findings
from the smaller experiments. The results of these
problem-oriented research efforts can inform policy about
appliance labeling more usefully than can predictions from
a model. They may also prove useful to modelers by providing parameter estimates that would not otherwise be
available in any empirically supported form. It might be
possible, for example, to interpret information on the
effect of labels as a change in a discount rate or a lag
coefficient. (In a discrete choice model equation such
as the one in Appendix A, labels might change the coefficient of response to energy efficiency.)

Sometimes research on qualitative factors such as program implementation or interpersonal communication cannot
be used to estimate the parameters of variables in models
because the variables are too hard to define and measure
precisely. In those instances, however, problem-oriented
studies can estimate the range of uncertainty for those
parameters and can offer explanations for the variation.

Problem-oriented studies and analyses of existing data
can also be used to test models. A model that can predict
the results of evaluation research, field experiments, or
analysis of RECS data is more likely to be correct than
one that cannot. Given the uncertainties in models, it
would be wise practice to compare the output and assumptions of models with empirical findings as a way of testing and refining models.

Thus, a multimethod approach would have several implications for energy modeling. It would lead to many small

changes in models--in parameter values and possibly in the ways variables such as price are represented--and it would probably change the variables included in models. And when qualitative factors, such as trust in information, prove important, it might lead to important innovations in the ways models are structured. In all these ways, energy modeling would be improved.

A multimethod approach to research would also affect the conduct of problem-oriented research. Models would help set priorities for other research by identifying unanticipated effects of policies that call for more specific attention and, when a model's output depends critically on the value of a particular parameter and the estimate of that parameter is uncertain, by calling for research or data collection, using other methods, to estimate that value.

The most important change that might arise from a multimethod approach, we hope, would be a shift of emphasis in the way energy demand analysis is conducted. Consumers of energy demand analysis might become less inclined to see in models the distillation of all knowledge about energy demand and more willing to see models realistically, as part of an ongoing process of analysis that relies on many techniques to build understanding.

References

Afriat, S.
  1973   On a system of inequalities in demand analysis:
          an extension of the classical method.
          International Economic Review 14:460-472.

Ajzen, I., and Fishbein, M.
  1977   Attitude-behavior relations: a theoretical
          evaluation and review of empirical research.
          Psychological Bulletin 84:888-918.

Archibald, R., and Gillingham, R.
  1978a   An Analysis of Consumer Demand for Gasoline
          Using Household Survey Data. Bureau of Labor
          Statistics, Washington, D.C., October.
  1978b   Consumer Demand for Gasoline: Evidence from
          Household Diary Data. Bureau of Labor
          Statistics, Washington, D.C., November.

Ascher, W.
  1978   Forecasting: An appraisal for policy-makers
          and planners. Baltimore: Johns Hopkins.

Atkinson, S. E.
  1977   Responsiveness to time-of-day electricity
          pricing: first empirical results. Part 1,
          pages 177-198, in A. Lawrence, ed., Forecasting
          and Modeling Time-of-day and Seasonal
          Electricity Demand. Palo Alto: Electric Power
          Research Institute.

Backus, G.
  1981   Preliminary Documentation for DEMAND'81 U. S.
          Energy Demand Policy Model. Vol. 1. School of
          Industrial Engineering, Purdue University.

Barnes, R., Gillingham, R., and Hagemann, R.
  1982   The short-run residential demand for natural
          gas. The Energy Journal 3(1):59-72.

Beck, P., Doctors, S. I., and Hammond, P. Y.
  1980   Individual Energy Conservation Behaviors.
          Cambridge, Mass: Oelgeschlager, Gunn & Hain.
Becker, L. J.
  1978   The joint effect of feedback and goal setting
          on performance: a field study of residential
          energy conservation. Journal of Applied
          Psychology 63:428-433.
Becker, L. J., Seligman, C., and Darley, J. M.
  1979   Psychological Strategies to Reduce Energy Con-
          sumption. Project summary report prepared for
          the U. S. Department of Energy. Center for
          Energy and Environmental Studies, Princeton
          University, N.J.
Berkovec, J., Hausman, J. A., and Rust, J.
  1983   Heating System and Appliance Choice. Working
          paper, Department of Economics, Massachusetts
          Institute of Technology, January.
Berry, L. G.
  1981   Review of Evaluations of Utility Home Energy
          Audit Programs. ORNL/CON-58. Oak Ridge,
          Tenn.: Oak Ridge National Laboratory.
  1982   The Role of Financial Incentives in Utility-
          Sponsored Residential Conservation Programs: A
          Review of Customer Surveys. ORNL/CON-102. Oak
          Ridge, Tenn.: Oak Ridge National Laboratory.
Blattenberger, G. R.
  1977   Block Rate Pricing and the Residential Demand
          for Electricity. Unpublished doctoral disserta-
          tion, University of Michigan, Ann Arbor.
Bohi, D.
  1981   Analyzing demand behavior: A study of energy
          elasticities. Baltimore: Johns Hopkins.
Breiman, L., Friedman, J., Olshen, R., and Stone, C.
  1984   Tree Structured Methods in Classification and
          Regression. Belmont, Calif.: Wadsworth.
Brewer, G. D.
  1983   Some costs and consequences of large-scale
          social systems modeling. Behavioral Science
          28:166-185.
Brownstone, D.
  1980   An Econometric Model of Consumer Durable Choice
          and Utilization Rates. Unpublished doctoral
          dissertation, University of California,
          Berkeley.
Bryon, R. P.
  1970   Restricted Aitken estimation of sets of demand
          relations. Econometrica 38:816-830.

California Public Utilities Commission
  1980   Energy Efficiency and the Utilities: New
          Directions. Symposium held at Stanford
          University, April 18-19.
Campbell, D. T., and Stanley, J. C.
  1966   Experimental and Quasi-Experimental Designs for
          Research. (Reprinted from Handbook of Research
          on Teaching, 1963) Chicago: Rand-McNally.
Caves, D. W., and Christensen, L. R.
  1978   Econometric analysis of the Wisconsin TOU
          electricity experiment. In D. Aigner, ed.,
          Modeling and Analysis of Electricity Demand by
          Time-of-Day. Palo Alto, Calif.: Electric
          Power Research Institute.
Center for Renewable Resources
  1980   Renewable Resources: A National Catalog of
          Model Projects. (four volumes) Washington,
          D.C.: U. S. Department of Energy.
Chernoff, H.
  1983   Individual purchase criteria for energy-related
          durables: the misuse of life cycle cost. The
          Energy Journal 4(4):81-86.
Cook, T. D., and Campbell, D. T.
  1979   Quasi-Experimentation: Design and Analysis
          Issues for Field Settings. Boston: Houghton-
          Mifflin.
Court, R. H.
  1967   Utility maximization and the demand for New
          Zealand meats. Econometrica 35:424-446.
Dahl, C. A.
  1979   Consumer adjustment to a gasoline tax. Review
          of Economics and Statistics 61(3):427-432.
Darley, J. M., and Beniger, J. R.
  1981   Diffusion of energy-conserving innovations.
          Journal of Social Issues 37(2):150-171.
Deaton, A.
  1983   Demand analysis. In A. Griliches and M. D.
          Intrilligator, eds., Handbook of Econometrics.
          Amsterdam, The Netherlands: Elsevier.
Diewert, W. E.
  1973   Afriat and revealed preference theory. Review
          of Economic Studies 40:419-426.
Diewert, W. E.
  1974   Intertemporal consumer theory and the demand
          for durables. Econometrica 42:497-516.
Donoho, D., Huber, P., and Thoma, M.
  1981   The use of kinetic displays to represent high

dimensional data. In W. Eddy, ed. Computer Science and Statistics, Proceedings of the 14th Annual Symposium on the Interface.

Dubin, J. A.
1982   Economic Theory and Estimation of Demand for Consumer Durable Goods and Their Utilization: Appliance Choice and the Demand for Electricity. MIT-EL-82-035. Department of Economics, Massachusetts Institute of Technology.

Dubin, J. A., and McFadden, D.
1984   An econometric analysis of residential appliance holdings and consumption. Econometrica 52(2): 345-362.

Eichenbaum, M., and Hansen, L.
1983   Uncertainty Aggregation and the Dynamic Demand for Consumption Goods. Working paper, Department of Economics, University of Chicago.

Elgin, D.
1981   Voluntary simplicity. New York: Morrow.

Energy and Environmental Analysis
1980   Documentation for the New Highway Fuel Consumption Model. Prepared for the U. S. Department of Energy. Transportation Support Services Task 13, January.

Energy Information Administration
1979   Single-Family Households: Fuel Oil Inventories and Expenditures. National Interim Energy Consumption Survey. DOE/EIA-0207/1. Washington, D.C.: U. S. Department of Energy.

Energy Information Administration
1980   Residential Energy Conservation Survey: Conservation. DOE/EIA-0207/3. Washington, D.C.: U. S. Department of Energy.

Ester, P., and Winett, R. A.
1982   Toward more effective antecedent strategies for environmental programs. Journal of Environmental Systems 11:201-221.

Farhar-Pilgrim, B., and Unseld, C.
1982   America's Solar Potential: A National Consumer Study. New York: Praeger.

Fisherkeller, M. A., Friedman, J. H., and Tukey, J. W. T.
1974   PRIM-9. An Interactive Multidimensional Data Display System. Stanford Linear Accelerator Publication No. 1408. Palo Alto, Calif.

Fitchburg Office of the Planning Coordinator
1980   Fitchburg Action to Conserve Energy (FACE) Final Report. Fitchburg Office of the Planning Coordinator, Massachusetts.

Freedman, D.
 1981   Some pitfalls in large econometric models: a
         case study. Journal of Business 54:479-500.
Freedman, D., Rothenberg, T., and Sutch, R.
 1983a   On energy policy models. Journal of Business
         and Economic Statistics 1:24-32.
 1983b   Rejoinder. Journal of Business and Economic
         Statistics 1:36.
Friedman, J., and Tukey, J. W. T.
 1974   A projection pursuit algorithm for exploratory
         data analysis. IEEE Transactions and Computers
         9:881-890.
Garevitch, M.
 1961   The Social Structure of Acquaintance Networks.
         Unpublished doctoral dissertation. Massachu-
         setts Institute of Technology.
Geller, E. S.
 1981   Evaluating energy conservation programs: Is
         verbal report enough? Journal of Consumer
         Research 8:331-335.
Geller, E. S., Winett, R. A., and Everett, P. B.
 1982   Preserving the Environment: New Strategies for
         Behavior Change. New York: Pergamon.
Goett, A. and McFadden, D.
 1982   Residential End-Use Energy Planning System
         (REEPS). Prepared for the Electric Power
         Research Institute, EPRI EA-2512. Cambridge
         Systematics, Inc., Massachusetts.
 1984   Residential end-use energy planning system:
         simulation model structure and empirical
         analysis. In J. R. Moroney, ed., Advances in
         the Economics of Energy and Resources. Palo
         Alto, Calif.: Electric Power Research
         Institute.
Greenberger, M., Crenson, M. A., and Crissey, B. L.
 1976   Models in the policy process: Public decision
         making in the computer age. New York: Russell
         Sage Foundation.
Greene, D. L., Hirst, E., Soderstrom, J., and Trimble, J.
 1982   Estimating the Total Impact on Energy Consump-
         tion of Department of Energy Conservation
         Programs. ORNL-5925. Oak Ridge, Tenn.: Oak
         Ridge National Laboratory.
Griliches, Z.
 1967   Distributed lags: a survey. Econometrica
         36:16-49.

Hausman, J. A.
 1979   Individual discount rates and the purchase and
          utilization of energy-using durables. Bell
          Journal of Economics 10:33-54.
Heberlein, T. A., and Warriner, G. K.
 1982   The Influence of Price and Attitude on Shifting
          Residential Electricity Consumption from On to
          Off-Peak Periods. Paper presented at the
          International Conference on Consumer Behaviour
          and Energy Policy, Noordwijkerhout, Netherlands,
          September.
Helson, H.
 1964   Adaptation Level Theory. New York: Harper &
          Row.
Hill, D. H.
 1978   Home production and the residential electric
          load curve. Resources and Energy 1(1):339-358.
 1983   The dynamics of private automobile transporta-
          tion demand. Unpublished manuscript. Survey
          Research Center, University of Michigan, Ann
          Arbor.
Hill, D. H., Groves, R., Howrey, E., Kline, A., Kohler,
 D., Lepkowski, J., and Smith, M.
 1978   Evaluation of the Federal Energy Administra-
          tion's Load Management and Rate Design
          Demonstration Projects. Palo Alto: Electric
          Power Research Institute.
Hill, D. H., Ott, D., Taylor, L., and Walker, J.
 1983   Incentive payments in time-of-day electricity
          pricing experiments: the Arizona experience.
          Review of Economics and Statistics 65:59-65.
Hirst, E. and Carney, J.
 1978   The ORNL Engineering-Economic Model of
          Residential Energy Use. ORNL/CON-24. Oak
          Ridge, Tenn.: Oak Ridge National Laboratory.
Hirst, E., Fulkerson, W., Carlsmith, R., and Wilbanks, T.
 1982   Improving energy efficiency: the effectiveness
          of government action. Energy Policy 10(2):
          131-142.
Hirst, E., and Goeltz, R.
 1983   Comparison of Actual and Predicted Energy
          Savings in Minnesota Gas-Heated Single-Family
          Homes. ORNL/CON-147. Oak Ridge, Tenn.: Oak
          Ridge National Laboratory.
Hirst, E., Goeltz, R., and Manning, H.
 1982   Household Retrofit Expenditures and the Federal
          Residential Energy Conservation Tax Credit.

ORNL/CON-95. Oak Ridge, Tenn.: Oak Ridge
National Laboratory.

Hirst, E., Goeltz, R., Thornsjo, M., and Sundin, D.
1983   Evaluation of Home Energy Audit and Retrofit
Loan Programs in Minnesota: The Northern
States Power Experience. ORNL/CON-136. Oak
Ridge, Tenn.: Oak Ridge National Laboratory.

Hirst, E., White, D., and Goeltz, R.
1983a   Comparison of Actual Electricity Savings with
Audit Predictions in the Bonneville Power
Administration Residential Weatherization Pilot
Program. ORNL/CON-142. Oak Ridge, Tenn.: Oak
Ridge National Laboratory.
1983b   Energy Savings Due to the Bonneville Power
Administration Residential Weatherization Pilot
Program Two Years After Participation.
ORNL/CON-146. Oak Ridge, Tenn.: Oak Ridge
National Laboratory.

Houthakker, H. S., Verleger, P. K., and Sheehan, D. P.
1974   Dynamic demand analysis for gasoline and
residential electricity. American Journal of
Agricultural Economics 56:412-418.

Huber, P.
1981   Projection Pursuit. Technical Report PJH-4,
Department of Statistics, Harvard University.

Kahneman, D., Slovic, P., and Tversky, A.
1982   Judgment Under Uncertainty: Heuristics and
Biases. Cambridge: Cambridge University Press.

Kahneman, D. and Tversky, A.
1979   Prospect theory: an analysis of decision making
under risk. Econometrica 47:263-291.

Kempton, W., Harris, C. K., Keith, J. G., and Weihl, J. S.
1982   Do Consumers Know "What Works" in Energy Conservation? Paper presented at the American
Council for an Energy-Efficient Economy Summer
Study, Santa Cruz, California, August.

Kempton, W., and Montgomery, L.
1982   Folk quantification of energy. Energy
7:817-827.

Klein, N.
1983   Utility and decision strategies: a second look
at the rational decision maker. Organizational
Behavior and Human Performance 31:1-25.

Koyck, L. M.
1954   Distributed Lags and Investment Analysis.
Amsterdam, The Netherlands: North-Holland.

Kwast, M. L.
  1980   A note on the structural stability of gasoline
          demand and the welfare economics of gasoline
          taxation. Southern Economic Journal 46:
          1212-1220.
Leonard-Barton, D.
  1980   The Role of Interpersonal Communication
          Networks in the Diffusion of Energy Conserving
          Practices and Technologies. Unpublished paper,
          Sloan School of Management, Massachusetts
          Institute of Technology.
  1981   Voluntary simplicity lifestyles and energy
          conservation. Journal of Consumer Research
          8:243-252.
Lerman, D. I., Bronfman, B. H., and Tonn, B.
  1983   Process Evaluation of the Bonneville Power
          Administration Residential Weatherization Pilot
          Program. ORNL/CON-138. Oak Ridge, Tenn.: Oak
          Ridge National Laboratory.
Luce, R. D.
  1977   The choice axiom after twenty years. Journal
          of Mathematical Psychology 15:215-233.
Marshall, N.
  1983   HID: A Dynamic Model of Change in the U. S.
          Housing Industry: Simulation of Long-Term
          Changes in the U. S. Housing Stock and in
          Residential Space Heat Demands. Report prepared
          for the Solar Energy Research Institute.
          Resource Policy Center, Dartmouth College,
          Hanover, N. H., January.
McFadden, D.
  1981   Economic models of probabilistic choice. Pages
          198-272 in C. Manski and D. McFadden, eds.,
          Structural Analysis of Discrete Data with
          Econometric Applications. Cambridge, Mass.:
          MIT Press.
  1983   Ten Years After the Oil Crisis: What Have We
          Learned? Paper Presented at Massachusetts
          Institute of Technology, November.
McFadden, D., and Dubin, J.
  1982   A Thermal Model for Single-Family Owner-
          Occupied Detached Dwellings in the National
          Interim Energy Consumption Survey. Revised
          manuscript, Department of Economics, Massachu-
          setts Institute of Technology, January.
McFadden, D., Puig, C., and Kirschner, D.
  1977   Determinants of the long-run demand for

electricity. Proceedings of the Business and Economic Statistics Section, American Statistical Association Part I:109-117.

McNees, S. K.
1979   The forecasting record for the 1970s. New England Economic Review September/October:33-53.

McNutt, B., and Rucker, E.
1981   Impact of Fuel Economy Information on New Car and Light Truck Buyers. Washington: U. S. Department of Energy.

Meier, A. K., and Whittier, J.
1983   Consumer discount rates implied by purchases of energy-efficient refrigerators. Energy 8(12): 957-962.

Mettler-Meibom, B., and Wichmann, B.
1982   The influence of information and attitudes toward energy conservation on behavior. (Translated by M. Stommel, Michigan State University.) In H. Schaefer, ed., Einfluss des Verbraucherverhaltens auf den Energiebedarf Privater Haushalte. Berlin: Springer-Verlag.

Mishkin, F.
1983   A Rational Expectations Approach to Macroeconomics. Chicago: University of Chicago Press.

Mosteller, F., and Mosteller, G.
1979   New statistical methods in public policy Part I: experimentation. Journal of Contemporary Business 8:79-92.

Nordin, J. A.
1976   A proposed modification of Taylor's demand analysis: comment. Bell Journal of Economics 7:719-721.

Northeast Utilities
1981   Consumer Survey of Potential Interest in a Conservation Loan Program. Prepared by Research for Policy Decision, Hartford, November.

Office of Technology Assessment
1980   Residential Energy Conservation. Montclair, N. J.: Allanheld, Osmun.
1982   Energy Efficiency of Buildings in Cities. Washington, D. C.: Office of Technology Assessment.

Olsen, M., and Cluett, C.
1979   Evaluation of the Seattle City Light Neighborhood Energy Conservation Program. Seattle, Wash.: Battelle Human Affairs Research Center.

Pacific Gas and Electric Company
 1982   ZIP Follow-up Survey. MR-82-1. Marylander
         Marketing Research, Encino, Calif.
Pallak, M. S., Cook, D. A., and Sullivan, J. J.
 1980   Commitment and energy conservation. In L.
         Bickman, ed., Applied Social Psychology
         Annual. Vol. 1. Beverly Hills: Sage
         Publications.
Pool, I., and Kochen, M.
 no   A Nonmathematical Introduction to a Mathe-
 date   matical Model. Unpublished manuscript.
         Massachusetts Institute of Technology,
         Cambridge, Mass.
Rogers, E. M., with Shoemaker, F.
 1971   The Communication of Innovations. New York:
         Free Press.
Rosenberg, M.
 1980   The RCS Project--Preliminary Findings and
         Recommendations. Technical Development
         Corporation, Boston, Mass.
Ross, M. H., and Williams, R. H.
 1981   Our Energy: Regaining Control. New York:
         McGraw-Hill.
Salama, S., Greene, J., and Krantzman, J.
 1980   End-Use Energy Demand Projections for Non-U. S.
         OECD Countries. Report prepared for U. S.
         Department of Energy by Energy and Environ-
         mental Analysis, Arlington, Va., August.
Samuelson, P. A.
 1970   Foundations of Economic Analysis. (2nd ed.)
         New York: Atheneum.
Sant, R. W.
 1979   The Least-Cost Energy Strategy. Pittsburgh,
         Pa.: Carnegie-Mellon University Press.
Sant, R. W., Carhart, S. C., with Bakke, D. W., and
Mulherkar, S. S.
 1981   Eight Great Energy Myths: The Least-Cost
         Energy Strategy--1978-2000. Energy Produc-
         tivity Report No. 4. The Energy Productivity
         Center, Mellon Institute, Arlington, Va.
Scherer, J. B.
 1981   Residential Energy Conservation and Financial
         Incentives. Prepared for the U. S. Department
         of Energy by Pacific Northwest Laboratory,
         Richland, Wash., June.
Shippee, G.
 1980   Energy consumption and conservation psychology:

a review and conceptual analysis. *Environmental Management* 4:297-314.

Solar Energy Research Institute
  1981   *A New Prosperity: Building a Sustainable Energy Future.* Andover, Mass.: Brick House.

Stern, P. C., and Aronson, E., eds.
  1984   *Energy Use: The Human Dimension.* Committee on Behavioral and Social Aspects of Energy Consumption and Production, National Research Council. New York: W. H. Freeman.

Stern, P. C., Black, J. B., and Elworth, J. T.
  1981   *Home Energy Conservation: Issues and Programs for the 1980s.* Mount Vernon, N. Y.: Consumers Union Foundation.

  1982a   Influences on Household Energy Adaptations. Paper presented to the American Association for the Advancement of Science, Washington, D. C.

  1982b   Personal and Contextual Influences on Household Energy Adaptations. Paper presented at the International Conference on Consumer Behaviour and Energy Policy, Noordwijkerhout, The Netherlands, September.

Stern, P. C., and Oskamp, S.
  1984   Managing scarce environmental resources. In D. Stokols and I. Altman, eds. *Handbook of environmental psychology*. New York: John Wiley.

Stobaugh, R., and Yergin, D., eds.
  1979   *Energy Future.* New York: Random House.

Svenson, O.
  1979   Process descriptions of decision making. *Organizational Behavior and Human Performance* 23:86-112.

Taylor, L.
  1978   The demand for energy: a survey of price and income elasticities. In W. D. Nordhaus, ed., *International Studies of Energy Demand.* Amsterdam, The Netherlands: North Holland.

Taylor, L., Blattenberger, G., and Verleger, P. K.
  1980   The Residential Demand for Energy. Palo Alto, Calif.: Electric Power Research Institute.

Theil, H.
  1965   Information approach to demand analysis. *Econometrica* 33:67-87.

Thurstone, L. L.
  1927   A law of comparative judgment. *Psychological Review* 34:273-286.

Travers, J., and Milgram, S.
1969   An experimental study of the small world
         problem. Sociometry 32:425-443.
Tversky, A.
1972   Elimination by aspects: a theory of choice.
         Psychological Review 79:281-299.
Tversky, A., and Kahneman, D.
1974   Judgment under uncertainty: heuristics and
         biases. Science 185:1124-1131.
1981   The framing of decisions and the psychology of
         choice. Science 211:453-458.
Tversky, A. and Sattath, S.
1979   Preference trees. Psychological Review 86:
         542-573.
Varian, H.
1982   Non-parametric methods in demand analysis.
         Economic Letters 9:23-29.
Winett, R. A., and Neale, M. S.
1979   Psychological framework for energy conservation
         in buildings: strategies, outcomes, directions.
         Energy and Buildings 2:101-116.
Winett, R. A., Hatcher, J. W., Fort, T. R., Leckliter, J.
N., Love, S. Q., Riley, A. W., and Fishback, J. F.
1982   The effects of videotape modeling and daily
         feedback on residential electricity
         conservation, home temperature and humidity,
         perceived comfort, and clothing worn: winter
         and summer. Journal of Applied Behavior
         Analysis 15: 381-402.
Winkler, R. C., and Winett, R. A.
1982   Behavioral interventions in resource manage-
         ment: a systems approach based on behavioral
         economics. American Psychologist 37:421-435.
Yankelovich, D.
1982   New Rules. New York: Bantam.
Yates, S.
1982   Using Prospect Theory to Create Persuasive
         Communications About Solar Water Heaters and
         Insulation. Unpublished doctoral dissertation,
         University of California, Santa Cruz.
Zarnowitz, V.
1979   An analysis of annual and multiperiod quarterly
         forecasts of aggregate income, output, and the
         price level. Journal of Business 52:1-34.

Appendix A
Incorporating Data from Problem-Centered Research on Incentives and Information in Formal Demand Models

Problem-oriented studies can investigate the effect of the size and type of incentives, of qualitative factors in the context of the incentive, and of different ways of offering information about energy efficiency. When such studies produce quantitative results, those results can be incorporated into formal energy demand models. This appendix presents a method for using such data in discrete choice models.

The typical state-of-the-art residential energy model includes equations that predict household purchases of new energy-efficient systems (appliances, heating equipment, retrofit of existing homes, thermal performance of new homes). Typically, these are discrete choice models to reflect the finite number of options consumers face for each class of purchase decision. For example, a model may define eight choices of water heaters (the combinations of two size classes, two fuels, and two levels of insulation). The choice models that predict what fraction of each equipment type is purchased might be of the form:

$$\ln(S_i/S_1) = a_0 + a_1 p_e U_i/eff_i + a_2 C_i + \cdots$$
$$- a_1 p_e U_1/eff_1 - a_2 C_1 \, ,$$

where S_i is the fractional share of new water heaters sold that are of type i, S_1 is the fractional share of type 1 water heaters, p_e is the price of electricity, U_i is annual hours of use of type i water heaters, eff_i is the energy efficiency of type i water heaters, C_i is the cost to the consumer of type i water heaters, U_1 is the annual hours of use of type 1 water heaters, eff_1 is the energy efficiency of type 1 water heaters and C_1 is the cost of type 1 water heaters. The coefficients a_1 and a_2 represent the weights of operating and capital costs in the choice; together, they determine the implicit discount rate for energy efficiency in purchases of new water heaters.[1]

Predicted values of equipment purchases are used to project stocks of energy-using equipment. These stocks are then used in conditional demand models to estimate household energy demand.[2]

Studies of variables that affect appliance choice yield findings that can be used to adjust either the coefficient a_2 or the equipment cost C_i in the purchase choice models. For example, information on consumer preference among different levels and types of incentives, stratified by type of consumer (e.g., by household income or education of household head) can be used to determine indifference curves, along which households are indifferent to choices such as those between rebates and loans (Figure A-1). These indifference curves can be converted to implicit discount rates to yield equations of the form:

$$a_2 = f(\text{incentive type and level, household characteristics}).$$

Such equations, which may vary with the type of purchase decision, can be used to determine the coefficient a_2, which can then be used in the purchase choice model.

Alternatively, the same kinds of findings might be used to estimate a parameter a_3 for some dummy variable, such as rebate versus loan or presence of an

---

[1]If d is the implicit discount rate for choices among water heaters with different efficiency levels, g is the consumer's expectation of future growth rates in fuel price, L is the expected lifetime of the water heater, and U is the expected annual hours of water heater usage, then the relationship between the coefficients and d is

$$a_2/a_1 = (d - g)/U[1 - e^{-(d-g)L}] \ .$$

If g is close to zero and L is large, then $a_2/a_1 = d/U$.
[2]The present discussion does not incorporate possible effects of appliance efficiency on U, subsequent appliance usage. In practice, purchase of a more efficient piece of equipment is likely to lead to greater use of that equipment than would occur with purchase of a less efficient unit. This occurs because higher efficiency leads to lower operating costs.

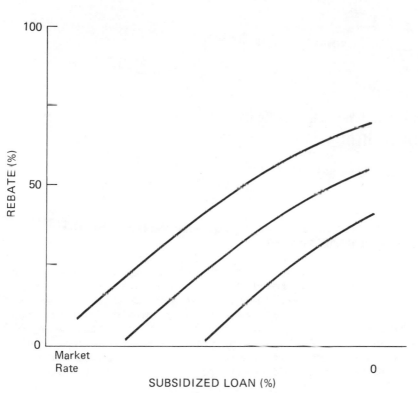

FIGURE A-1 Indifference curves for loans versus rebates
for different groups of consumers, e.g., stratified by
income.

appliance label of a certain design. This procedure would
be justifiable under an assumption that a particular dummy
variable affects choice additively, rather than as a multiple of appliance cost. Still other interpretations are
theoretically plausible. For example, appliance labels
may call attention primarily to the operating cost of the
appliance. This effect could be modeled by using data on
actual choices to estimate a_1. Of course, the adequacy
of one or another of the plausible interpretations of data
is an empirical question.

Appendix B
Biographical Sketches of Panel Members and Staff

JOHN M. DARLEY is a professor and chairman of the Department of Psychology at Princeton University; he previously taught at New York University. He currently studies perceptions of energy and energy problems and the ways in which information can be made available to people to facilitate their energy-conserving behaviors. His earlier work included research on people's reactions to emergencies, particularly reactions that determine whether people will give help to victims. He received a B.A. degree from Swarthmore College and a Ph.D. degree in social relations and social psychology from Harvard University.

DAVID A. FREEDMAN is a professor of statistics and mathematics and chairman of the Department of Statistics at the University of California, Berkeley. His recent research concerns the validity of statistical models used in the social and behavioral sciences; the behavior of standard statistical procedures under nonstandard assumptions; and the evaluation of econometric models, including those used in energy policy analysis. He has a B.Sc. degree from McGill University and M.A. and Ph.D. degrees from Princeton University.

DANIEL H. HILL is an assistant research scientist at the Institute for Social Research, University of Michigan. His research interests are in survey methodology, analysis of gasoline demand and time-of-day electricity consumption, and the environmental impacts of transportation, and he has worked for the past ten years on the Panel Survey of Income Dynamics. He received a B.A. degree from the University of North Carolina at

Chapel Hill and M.A. and Ph.D. degrees in economics from the University of Michigan at Ann Arbor.

ERIC HIRST is a group leader in the Energy Division of Oak Ridge National Laboratory; he previously worked at the Federal Energy Administration and the Minnesota Energy Agency. His current research involves quantitative evaluation of utility and government energy conservation programs; in the 1970s, he developed and applied the Oak Ridge National Laboratory's residential and commercial energy use models. He received a B.M.E. degree from Rensselaer Polytechnic Institute and M.S. and Ph.D. degrees in mechanical engineering from Stanford University.

DANIEL McFADDEN is James R. Killian professor of economics at the Massachusetts Institute of Technology; he was previously a professor of economics at the University of California, Berkeley. His research is in econometrics and economic theory with applications in energy, transportation, and the demand for consumer durables. He is a member of the National Academy of Sciences and the American Academy of Arts and Sciences and president-elect of the Econometrics Society. He received his B.S. and Ph.D. degrees in economics from the University of Minnesota.

LINCOLN E. MOSES is a professor in the departments of statistics and of preventive medicine at Stanford University. Previously he was dean of graduate studies and associate dean of humanities at Stanford. From 1978 to 1980 he served as the first administrator of the Energy Information Administration in the U. S. Department of Energy. His principal interests are in the applications of statistics to medical and behavioral research. He is a member of the Institute of Medicine and a fellow of the American Statistical Association and the Institute of Mathematical Statistics. He received an A.B. degree from Dartmouth College and a Ph.D. degree in sociology from Brown University.

PAUL C. STERN is study director for the Committee on Behavioral and Social Aspects of Energy Consumption and Production and its Panel on Energy Demand Analysis. Previously he was research associate in the Program on Energy and Behavior at Yale University's Institution for Social and Policy Studies. His research interests are in

energy use and environmental policy issues. He heads the
energy committee of the Division of Population and
Environmental Psychology of the American Psychological
Association. He has a B.A. degree from Amherst College
and a Ph.D. degree in psychology from Clark University.